Take Heart
Philip D. Houck M.D.

Prologue: Not everything presented in this story of the heart is correct. Not everything I was taught about the heart is correct. The information in this story is what I believe to be correct; and hopefully, this information will improve your life and health and spirit.

Certificate of Registration

This Certificate issued under the seal of the Copyright Office in accordance with title 17, *United States Code*, attests that registration has been made for the work identified below. The information on this certificate has been made a part of the Copyright Office records.

Marybeth Peters

Register of Copyrights, United States of America

Registration Number:

TXu 1-572-371

Effective date of registration:

March 19, 2008

Chapter 1
Congestive Heart Failure

Symptoms of Heart Failure

Murray, beaten down by life, was older than his stated age of 82 years. His breathing had become harsh and measured. The weight on his lungs was ever constant. Youthful memories of the first race of the spring would bring a similar heaviness and tightness. In his youth, however, the feeling would fade away quickly. Now, each measured step is taken by counting his breaths. As his breaths become more frequent and more constrained, his steps slow down. As he stops to catch his breath, his head falls upon his chest with labored breathing. It is a taxing effort to pull more air into his aching lungs.

His hand tattooed by blue-black hemorrhages, and aged by many years of sunshine and hard labor, carries a 24-ounce cup. In this cup, his constant companion, there is a cool fluid to quench his never-ending thirst. Murray remembers from grade school that he should drink 8 glasses of water every day for good health. He has tripled this intake in the last three months. His body which is 90 percent water has gradually grown by 25 pounds which is equivalent to 3 gallons of excess water. Just like a water balloon, the fluid tends to float to his legs; making his shoes very tight. However, he is not prone to getting on the bathroom scales and is unaware his weight had increased 25 pounds. The fluid is in all of his tissues and has leaked into his abdomen making his belly tense. His pants are tight and he no longer buttons them but uses a belt to anchor the pants. The pressure of the fluid which Murray interprets as gas and bloating makes it difficult to breathe.

The first two months of this phenomenon were of no concern because by the next morning his puffy feet would be gone. As the oppressive day would grow, so would his legs and feet. In the third month he started to wear bedroom slippers to accommodate his enormous swollen feet. His concern began grow when he realized the edema would not leave his legs, even while he slept. If it can be achieved, sleep is a wonderful thing, however, the nighttime of renewal for Murray was unattainable.

He remembers when his head used to hit the pillow and would not move until the alarm would go off. Now he can sit and watch the clock strike every mournful hour of the never-ending evening. He counts the number of times that he arises to empty his painfully full bladder. He finds the first hour of sleep goes by quickly. Sleep is broken when his lungs begin to fill like a porous sponge. The first thing that is noted is a tickly cough. With each cough his breathlessness becomes more apparent. His lungs are burning and heavy. An overwhelming sense of panic begins to develop and increases his heart rate. He must rise, sit up, get to the window, and open the sticky and squeaky sash. The cool night air flows into the room and he leans forward to catch ever-deepening breaths. His panic begins to subside. He returns to his bed, and again tries to lie down. Within several minutes the cycle repeats itself. Murray goes to his favorite recliner and sleeps in a sitting position the rest of the night. The next day he feels fatigued and the day becomes sheer drudgery.

Wednesday began as usual with two cups of coffee, scrambled eggs, bacon, and toast. The glistening white crystals of salt fell from a shaker onto the eggs. The salt lit up the tongue with flavor obliterating the taste of the eggs. Salt is ever present hidden in processed foods thus becoming impossible to avoid. It is impossible not to swallow hidden salt in the diet. Restaurants continue to use the sensory trick of tainting food with salt to make the cuisine more palatable to their customers. Although salt is essential to daily nutrient requirement, the body only needs the small quantity of 500 mg which is nearly 10 times less than the average American diet. The cornflakes consumed at breakfast will meet the entire daily requirement.

In past times, salt was more essential in the diet. The quotation, "you're not worth your salt", is from the times of the Roman soldiers. The soldiers were paid in salt for their work. Bad soldiers were not deserving of their pay. Salt was necessary because the meat that they carried with them was often rancid and required the salty flavor in order to swallow the decaying meat. Refrigeration has been present for nearly a century. Meat is hardly ever rancid and rancid meat furthermore, is never eaten. However, the habit of salting our food has remained for the last two millennia. Salt, the only edible rock, is precious, omnipresent, and nearly impossible to avoid. The friend of the inexperienced chef is salt, and the foe of any hypertensive patron.

Salt is like a sponge. It requires water as a companion. In water it dissociates into sodium and chloride. The chemistry of how much water it takes to dissociate the ions and allow them to dissolve is covered in high school chemistry. The body has to maintain a certain concentration of salt. If excess salt is swallowed, then extra fluid must also be brought into the body to maintain a stable concentration of sodium 140 Meq. The end result is extra salt requires extra water and our sausage bodies swell to accommodate the excess water. The sodium concentration in blood one hundred forty milliequivalents is similar to seawater. According to Darwin, when we crawled out of the ocean the earth was 90% water just like the biologic beings that walked on its surface. Our primordial cells need the same concentration to survive. What allowed evolution to walk onto the land? Was it the mutation that allowed the lungs to develop?

The kidney was more vital to conquering the land than lungs. The kidney was to preserve fluid and only excrete the fluid necessary to carry the acid and breakdown products of metabolism. Other kidney functions include preservation of every drop of moisture and maintenance of blood pressure. Survival of the fittest and the principle of the food chain placed humans in a potential dangerous position. As an organism in the food chain, it would not be abnormal to suffer an injury and blood loss. The kidney has the power to constrict blood vessels, raise blood pressure and hold onto salt and water. These homeostatic mechanisms helped assure that mammals would survive until the next deadly comet. A failing kidney is unable to excrete salt. It will take a sick kidney several weeks to eliminate the salt consumed in the normal diet during one day. A little salt excess will cause pain and suffering in the form of breathlessness and swollen legs for weeks after the salt intake. The kidney, like a miser, would hold onto salt and the salt would act like a sponge; soon Murray's shoes would not fit even with the shoelaces removed.

As the lunch hour approached, breathing was becoming more forced and labored. Murray did not feel very hungry. He had a feeling of dread like turkey vultures flapping their black shroud wings. For lunch he had condensed soup thinking this would improve his spirits. Three glasses of water were required to quench his thirst and bring into balance the excess sodium that was in the soup. Murray was having trouble getting from one chair in his house to the next. His heart rate began to race, his blood pressure began to rise; and with each increase in his heart rate and each mmHg rise in his blood pressure, Murray became more panicked.

The next morning, he developed tingling in his lips and fingers and ever increasing numbness that was the result of his hyperventilation. His breath turned frothy and tinged with blood. In a panic 911 was called and an ambulance was summoned. By the time the paramedics arrived Murray was no longer thinking. He was hallucinating, combative, and struggling just as a drowning man in the middle of the ocean.

Flashback During Near Death Experience
During his heart failure delusions Murray's thought fell back in time and westward into the Asian Pacific. He was on a tropical island far from Manila. It was beautiful with mountains jumping to the sky in jagged peaks. They were so steep that areas of landslides are common. In the dry season, the slopes were golden brown at the higher altitudes and deep green at the lower levels. The pine trees constantly hum with a low pitch tune like a violin being stroked by the wind.

The time when Murray first arrived in the Philippines was not pleasant. He was part of a special operation team whose function was to disrupt the Japanese occupation. His on the job training was shear wit and intelligence. Twenty years later he would instruct US Special Forces in the techniques that he found successful. The knowledge that he gained in the Philippines, living off the land, independent from supply lines would eventually become standard operations in any conflict. The principle of covert operations, gaining the trust of local people and causing disruption in enemy routine operations is still vital in world conflicts. His only precious resource was a radio that he hid and would return to make updates on troop strength and movement. He was sent as an observer but transitioned to overt operations. His favorite disruption involved an illegal acquisition printing plates of Japanese currency. It had been a stroke of luck when a small band of Japanese curriers traveled into his ambush. Instead of transporting money from the homeland the Japanese would print their own money to acquire local goods and support. The currency plates were part of every company. Later in the conflict, offering printed cash was omitted and the army acquired by force their provisions of food, women, and slave labor. The printing plates were in a heavy chest that opened spontaneously and spilled to the ground as Murray's ambush neutralized the enemy combatants. The plates were now in the possession of Murray to devise dirty tricks.

With the help of the Negritos, a fearsome pygmy tribe, he printed money and used an old washing machine to age the currency. The Negritos who did not even own pants helped to spread the fake dirty money among the peasants. They were unaware the value of money since they had no economy and only lived off the land. Soon the large amount of currency circulated by the local people completely inflated the Japanese money making the real Japanese currency worthless. No one wanted to trade for this currency and in a small part the counterfeit money undermined the credibility of the Japanese among the native people. Unfortunately, any legitimate business was now impossible and nobody was paid.

The Negritos unfortunately realized that they did not have pants. Society turned their idyllist nature into one of searching for material goods. It would take 3 years of labor for the Negritos to make enough money to buy pants. Their favorite occupation was security. They would hide in the trees around homes they protected and surprise thieves. One ear from a thief was traded for a bounty that could help make a pants payment. Life was simple and good when there was no money and no pants.

Murray's travel drew him to the mountains near Baguio; past a great lion head rock into hills that had been sculpted by ancient engineers building terraced landscapes that could produce crops even on the steep slopes of the mountains. The terraces, stone bridges across gorges, survived for centuries. The civilization was a marvel, but as most ancients, slowly lost their identity to invaders, disasters, and disappeared leaving only the engineering marvels without the innovative people who designed them. Mining and the removal of metals was a common practice. There were Legends of great underground halls filled with fine silver crafted into jewelry, animals, dinnerware, and statues. Murray's travel would lead him into an ancient secret protected by mountains and shear inaccessibility.

He had been following a small scout expedition of Japanese engineers who were seeking high ground to monitor for US airman trying to soften the island defenses with napalm and TNT. They were walking thru a gorge when they spotted Murray following them. Concurrently Murray had sensed he had been detected and he began to climb a small outcropping. He had traveled two thirds of the way up the rocky escarpment when a grenade fell just short of his position. He fell before the grenade went off and landed on a narrow ledge. The grenade exploded and sent a shower of small stones dust onto his position. The Japanese had witnessed his fall and initially tried to reach his position to confirm the kill. After one near injury they abandoned the quest and continued with a bit more attention to the rear. When Murray awoke he saw a small opening in the side of the hill. Fearing that his liver would be eaten he slid into the crevice so he would not be found. To his surprise the crevice began to widen and the further he went the larger the opening became. He had stumbled into the entrance of a small cave or lava tube that leads into the mountain. He continued to move slowly hoping he would elude the Japanese. For three hours he waited listening for signs of pursuit. Cautiously he lit a candle he always carried and to his amazement he found a brick laden wall and not a natural cave. As he followed the wall on a well-trodden path he noticed small silver statues.

Initial Emergency Care and Physical Examination of Heart Failure

Murray's delusion faded as the paramedics applied oxygen, sublingual nitroglycerin, Lasix, and morphine. A Continuous Positive Airway Pressure (CPAP) Machine was fitted to his face. This machine would push pressurized air into his lungs to counter the edema fluid that was leaking into his lung air spaces. Initially Murray became panicked as he was breathing through what felt like a closed suffocating space. He was not sure where he was. He was no longer suffocating in a small cave in the Philippines. Once the machine began to work; and the pressurized air helped push the fluid out of his lungs Murray began to feel better. He stopped fighting the mask and allowed it to do its work. He was loaded onto the ambulance and halfway to the hospital his bladder had already filled from the injection of Lasix that he was given. Murray was overwhelmed. Just twenty minutes earlier he began to recount life events, certain that death was near. The ease of his restored breathing pattern filled Murray with hope and the will to live another day.

The emergency room doctor who greeted Murray had a reassuring smile trimmed by a perfect shade of lipstick. If Murray arrived 15 minutes earlier, his hypoxic delusion would have identified her as one of God's finest angels. She seemed tall and slender as Murray gazed up from the gurney. Her eyes were the darkest raven eyes he had ever seen. Her flawless complexion appeared like a diamond bordered by shoulder length jet-black shiny hair that moved in opposite light flashing directions as she quickly moved her head to assess Murray's shell of a body. The doctor observed his chest rising and falling, each with an expansive breath, 29 times per minute. She saw his pink tinged shirt from his blood-tinged sputum. The automatic blood pressure machine recorded a blood pressure of 170/100 and a variable pulse of 125 beats per minute. She then knew Murray would get better quickly.

She turned to the nurse before the rest of her examination and ordered "Lasix 40 mg IV, morphine 2 mg. IV, Please give a bolus of *Natrecor* 2umg/kg and an infusion at .01 mcg/kg/min".

Turning her perfectly sculptured chin toward Murray, she asked, "Do you use salt? Are you thirsty and frequently carry a container of water with you during the day?"

Murray answered with difficulty through his mask, "I just use a pinch now and then and I drink about the same as anybody else."

The dazzling and insightful doctor quipped, "Do you have a salt shaker? Ever eat chips, lunchmeat, and sauerkraut? Did you have chicken noodle soup for lunch or should I send that dangling noodle that is next to your chin to the parasite laboratory?" As she awaited the expected sheepish answer, she observed the jugular vein deep to the sternocleidomastoid muscle bellies. The external jugular was full but this could be a misrepresentation of the patient's condition. Even with Murray bolt upright she could see the top of the pulsating column of the internal jugular 15 centimeters above the right heart. This would correlate with the right heart being unable to push blood through the lungs. When Murray was standing 5 ft 8 inches tall, the venous pressure would exert tremendous pressure. This pressure would cause a slow leak of fluid into the peripheral tissues. During the night the venous column fell to less than 15 centimeters allowing the fluid to seep back into the blood vessels overfilling the capacity of the veins. The right heart would fill the lungs and the left atrial pressure would rise. The lungs would become heavy with the extra weight of the fluid. The work of breathing would rise with each new ounce of fluid. Sitting up or standing would move the fluid back to the legs to give the lungs a rest.

She knew Murray had been sleeping in a chair and therefore almost seemed unnecessary to ask, "Do you awaken with shortness of breath, smothering, and have to sit up? How many pillows do you use at night, or do you sleep in a recliner?" As she awaited the answer, she let her delicate fingers find the carotid artery located lateral to the voice box nestled against the sternoclavicular muscle body. The carotid was weak difficult to feel suggesting the amount of blood thru this artery to the brain was diminished. This finding correlated with Murray's skin being cool damp and untouchable. The cool damp and untouchable skin triggered a distant memory of a teenage boy who took three acts of a four acts play to reach over and hold her hand. This should have been pleasant but the nervous perspiration and nearly bloodless hand was uncomfortable almost corpse-like. Murray's cardiac output, the blood pumping ability of the heart was not love struck but markedly diminished. The output was so poor it could barely fill the vital carotid artery.

Forcing herself to forget her first teenage disappointment, the young doctor redirected her attention to Murray's chest. She let her fingers fall onto the chest just left of the sternum, feeling between the rib spaces. Just below her fingers pushing against the chest wall was the main artery to the lungs. The pulmonary artery was normally a silent inhabitant acting as a low-pressure conduit to the lungs. With congestive failure, the pressure would build into the lungs and back into the pulmonary artery. This high pressure would amplify the pulse coming from the right heart. A gentile, barely perceivable pulse could be felt between the ribs. The finding confirmed the pulmonary pressures were elevated far above normal and this meant serious illness. It is strange that such a specific finding was not taught in medical school. A gray haired, George Clooney handsome cardiologist first demonstrated this finding to the young ER doctor diagnosing mitral stenosis without a stethoscope. Her own pulse quickened as he touched her hand and placed it between the cachetic patient ribs. At first she felt nothing, then, during expiration a barely perceivable tap became evident. Once palpated it became intensely obvious. The following hour brought countless medical students and residents to the bedside of the patient with mitral stenosis.

The PMI (Point of Maximum Impulse) was the next objective in Murray's physical exam. The gray-haired cardiologist had instructed the young doctor 'not everything she had been taught in medicine was correct'. It was her job to question and understand every thing she is taught. The textbooks have instructed that the patient should be lying in the left lateral decubitus position to feel the PMI. It certainly is true that the PMI is easier to feel in this position. The heart falls towards the chest wall and the cardiac impulse travels through the chest wall. This position however does not provide a precise insight to the true strength and size of the heart. The correct position is in the sitting position with heart resting in its normal position. Many times in normal individuals the PMI cannot be palpated. However, if the diseased heart has become enlarged it will take up more of the chest cavity and will bounce farther into the lateral chest wall. After years of palpating this impulse, the gray-haired cardiologist could estimate the pumping ability of the heart and if the muscle was hypertrophied, thickened, or very weak. In Murray's case, the PMI was displaced and diffuse. This clue would suggest that the left ventricle is dilated, with weakened contraction. If one could imagine seeing into the chest just as the echocardiographic machine, one would see an enlarged spherical heart with trembling contractions barely perceivable. The mitral valve leaflet would not touch, coapt, allowing a torrent of blood to flood in the wrong direction back into the lungs.

The normal ejection fraction is not 100 percent. The ejection fraction measures the amount of blood that is expelled from heart. Normal hearts expel 60 percent of blood with forty percent remaining for the next contraction. The heart walls cannot collapse upon each other allowing for efficient flow through the heart. A hyperdynamic heart with an ejection fraction of seventy-five percent can cause syncope, a passing out spell. The opossum reflex of playing dead is an example of this reflex. Having the heart stop was a good way of preventing blood loss. Many people still exhibit this primitive reflex when they go get their blood drawn. When they see their own blood the heart slows blood pressure falls and a pass out is sure to occur unless the heart gets closer to the ground. The reflex can also be elicited when a hyperdynamic small heart gets smaller. With standing there is a tendency for the blood to pool in the feet and not return to the heart. People with hyperdynamic hearts can have their ejection fraction go even higher when they stand in Wal-Mart, or in church, or in a lunch line at Luby's. The heart becomes smaller and the reflex makes the heart slow down to improve filling. If you fall over in a faint and lay prostrate on the ground, the heart will get larger and recover. This is good for the heart but bad for social skills when people are expected to stay upright. Helpful bystanders will aggravate the condition by helping the affected person back to their feet where the heart will again empty and cause a recurrent syncope. When someone faints, it is beneficial for the heart to stay on the ground. Social skills should be secondary to the happiness of the heart.

A normal heart looks like a football. The giraffe, which has to pump blood many feet above the ground, has a very elongated heart to achieve the efficiency needed to pump to that great height. Patients with heart failure like Murray have spherical shaped hearts. These failed hearts are similar to cold-blooded animals that have very low blood pressure and spherical hearts. Alligators don't get on two legs and chase you. They do not have enough blood pressure to stand up. Murray's blood pressure has to be low, and he may benefit by taking an alligator stance with his belly on the ground instead of standing upright. The George Clooney of cardiology often joked to his students that patients with systolic dysfunction should be relegated to all fours and should slither around like an alligator. There should be special wheelchairs that allow patients to lie on their stomachs and propel themselves just gliding above the surface of the ground. Murray's heart is spherical and his blood pressure needs to be low for his heart to survive. Murray feels fatigue as a manifestation of low blood pressure. In time his body will adapt to the low blood pressure.

The dark raven-eyed emergency room doctor slipped her hand just under Murray's breast. There was an impact on the chest wall that was gentle but definitely perceivable. The size of the impact was 4 cm. The location of the impact was greater than three cm beyond Murray's nipple. Murray's ejection fraction was estimated at 20%. The emergency room doctor lifted her stethoscope that had been hanging around her neck like a necklace. The triple head stethoscope was a Tycos Harvey and was calibrated in length to her teacher's. Although not electronic, it could accurately reproduce sounds that were transmitted from the heart to the chest wall.

There was a time in the early seventeen hundreds that listening to a lady of the court's breast would result in amputation of the physician's ears or worse. Dr. Laennec, the inventor of the stethoscope, had found that by rolling a paper into a tube he could discreetly listen to the chest without offending the sensibilities of the court. He later produced a wooden cylinder that could be marketed to other physicians. Heart sounds, lung sounds, and borrigismus (stomach tinkles) could now be heard improving the accuracy of diagnosis.

The *Estee Lauder* scented emergency room doctor was thankful that she could always keep some distance between herself and her more disagreeable and odiferous patients. Pressing the stethoscope to Murray's back, the doctor could hear crackles and wheezes in the lungs. The proper term was rales, and rhonchi; indicating pulmonary edema was present. Some of the 25 pounds of excess fluid was overflowing into the airways making breathing more difficult.

While listening to the neck, she could hear a soft vibration in the left carotid indicating blockage of blood flow to the brain. Cholesterol buildup into the carotid artery would eventually close the vessel resulting in a possible stroke. She knew, however, this was only a marker and what Murray would most likely die of is a heart attack. Bruits, the vibration in the neck means vascular disease is present. When this occurs in one vascular distribution like the neck vessels to the brain, it is likely to be in all distributions including the heart, the aorta, and leg vessels. The head, the heart, and the legs take the brunt of the symptoms. Having a bruit is a marker for a future heart attack. The other simple screen for atherosclerosis (cholesterol and inflammatory build up inside the blood vessels) was to take the blood pressure in the ankle. Systolic blood pressure rises as it travels down the aorta into the legs and feet. This is due to reflected waves of pressure bouncing off the blood vessel branches. The mean blood pressure falls as the vessel get farther from the heart. In patients with vascular disease the blockage causes the pressure to fall more quickly. This simple test is easily done and is an excellent predictor of patients who need aggressive therapy. If the pressure is lower in the ankle than in the arm the disease is present.

The doctor now directed her attention to the aortic area on the chest wall. She could hear a soft murmur that became louder as she placed the stethoscope closer to the left lateral chest. In the left lateral chest she could hear a holosystolic murmur, a blowing shssssss during contraction of the heart. This murmur is consistent with mitral regurgitation. She asked Murray to squeeze both hands into fists. By performing this maneuver it becomes more difficult for the blood to pass into the body and more will leak backwards to the left atrium through the leaking mitral valve. Murray's murmur increased with handgrip confirming the presence of mitral regurgitation. At the apex of the heart near the breast, the doctor placed the bell of the stethoscope and listened for vibration that was so low in pitch it could only be felt. The pitch occurred in the cadence indicating a gallop rhythm. The S3 indicated the ventricle was distended.

In summary, Murray was a bloated salt and fluid filled human with elevated jugular veins indicating right heart failure and fluid overload. His carotids were diminished indicating poor cardiac output and poor prognosis. The tap of his pulmonary artery indicated pulmonary hypertension and a high likelihood that he had mitral regurgitation, a leak of his mitral valve. This leak occurred because his ventricle was dilated stretching the mitral valve apparatuses beyond its normal size. His ejection fraction was estimated to be 20 percent. Murray's chance of dying in the next year was 20%, 50% in the next 5 years. He would have a greater chance of survival if he had cancer.

Hospitalization

Murray really did not remember much from the rest of the day. He had been bedded down in a two-person room. His nurse and nurse's aid were of very pleasant demeanor. After asking an appropriate number of questions, Murray was left to his wandering thoughts and the loud snores of his roommate.

Murray was now immersed in a new and confined life style. One was the tube placed into his bladder. The tube caused uncomfortable irritation with the feeling he had to urinate. He felt he had to get up to use the restroom, however, the nurse's aid repeatedly told him that he didn't need to get out of bed. The Foley Catheter would collect the urine for him. Beverly, the nurse's aid, was mildly plump with a friendly and caring smile. The Foley bag was now full of nearly clear liquid. Beverly came to drain the bag of a measured 750 cc of urine. The art of compassion was learned through many years of service to patients so difficult even a mother would refuse to love and care for. Beverly had two grown children, one of which had returned after a failed marriage. As a mother and a grandmother who experienced personal tragedy, she had learned to keep faith and meet challenges with a smile and hopeful expression. This optimistic attitude with her patients hastened their recovery.

Murray's nose was congested, dry and irritated. He had a canula stuck between both nostrils. They were arranged at the side of his face instead of in his nose. The oxygen was not humidified and tended to cause congestion. Pure oxygen was very dry and often would cause the nasal passages to swell from the dryness. The beds had improved a great deal from the cots of World War II but were still unfamiliar. The monitoring wires could entangle and choke and were like a leash until the space program and Duracell batteries allowed some minor freedom. On his skin were multiple round patches with metal nipples some of which had wires that entangled his arms. A call button was attached to the bedrail near the confusing controls for the bed and television. Despite the unfamiliar surroundings, Murray fell asleep and rested better than he had in the last two weeks. He awoke the next morning just prior to morning rounds.

After breakfast, an entourage of doctors, students, and residents burst into his room. The George Clooney of cardiology greeted Murray with a confident and reassuring smile. He said he was an old Air Force doctor and had noted that Murray was a ground pounder from the army. Following the Houckster was the emergency room beauty that was rotating into cardiology from the emergency room as part of her training schedule. Emergency doctors needed cardiology training because the most common mistakes were usually made in patients with heart disease. Most of the residents assumed the nickname Houckster came from Hulk Hogan the "Hulkster." The name was actually rendered by an artistic, dirty blonde female airman who had been trained as an Echo Tech. The Airman was over talented for her rank as were many of the Airmen. Some even had advanced degrees and chose the Air Force rather than being drafted into the Viet Nam conflict. She had worked closely with the young captain and respected him because he respected and appreciated her talent. The Airman designed a "Save the Houckster" campaign when he had gotten orders to the Philippines. A Houckster was a cross between a mustached, horned rimmed glass wearing cardiologist nerd and a hamster. Dr. Houck allowed the vision of Hulk Hogan be applied mistakenly. It certainly commanded more respect than a hamster.

Since she knew Murray, she presented the case to the staff cardiologist who confirmed her physical findings and confirmed the therapy. Murray piped in that the therapy was perfect since he felt great and therefore needed his pants to go home and have a smoke. The wise staff doctor told Murray that pants are always taken to assure fewer escapes from the hospital. Very few patients want to be confined in the environment of nighttime chirps, beeps, pages, and roar of cleaning equipment reverberating around the high acoustic tiles. Hospitals were designed so the monitoring area would not include an outside place to smoke. Patients needed help to break the tobacco habit and the hospital was a successful place to start.

Pizza Lecture to Illustrate Why Smoking is BAD!
The staff doctor then turned to the residents and said, "Who wants to give the pizza lecture?" Pizza lecture? A common puzzled look and reticence spread across their young faces. Seeing there were no takers for the lecture, the well-muscled, mustached physician began. "This is not an anti-pizza lecture, but pizza is used to explain why you should quit smoking. Imagine a hot pizza taken out of the oven with a bottom crust that is crisp brown and possibly baked a bit too long. Instead of eating the cheese topped pizza, you begin to bend the crust making the center dip into the center and spring back up. Murray, what is going to happen to that pizza crust after you bend the crust 10 to 12 times?"

Murray hesitated, for a minute and then replied. "The crust will crack and the cheese is going to drip through and hit your pants."

The staff doctor turned to the residents and smiled "Correct. So, young doctors, how is this story like a heart attack?" Their eager faces all appeared a blank shade of pale. The paleness gave the impression that they were invisible and they should not be called on to answer. The staff cardiologist turned to Murray and asked him to explain.

Murray hesitated ensuring he had the attention of all the young doctors and then began his answer: "Imagine the inside of your coronary arteries having a cholesterol placque, a small bubble inside the artery that protrudes into the flowing blood. The fibrous cap that covers the liquid pool of cholesterol is like the pizza crust and the cholesterol is like cheese. Every time you take a puff on your cigarette the intense vasoconstriction of nicotine will cause the artery to squeeze on the pizza crust plaque. Nicotine is a very active drug and within seven second will hit your brain and heart, as well as cause the vessel to constrict and then relax. Once the plaque is ruptured, the thrombogenic cholesterol leaks out and forms a clot. The clot shuts off the blood supply to the muscle supplied by the artery and the muscle dies. Smoking is an immediate risk for having a heart attack. It causes plaques to rupture. Throwing away your cigarettes will immediately lower your risk for a heart attack." Murray slumped back with a satisfied look on his face and the residents looked at each other in a dumbfounded astonished look. They had learned something from this worn out broken human.

The cardiologist then turned to Murray and responded, "Excellent, how long have you been off of your cigarettes?"

In a sheepish but defiant look Murray answered "since yesterday." This was not Murray's first visit to the hospital; he had been admitted several times for chest pain and threatened myocardial infarction. He only received the pizza lecture once, but with his photographic memory had committed and recited it word for word. Unfortunately, Murray did not feel that the pizza lecture applied to him. After all, he was invincible. He survived the Japanese occupation of the Philippines and numerous other adventures. His sense of being invincible, hard life, and many near encounters with death had left him with the feeling he could go on forever. He just wanted to tend his crops, animals, and do what he pleased. He now knew that he had to change. Something had changed. He was lacking his health. Murray now felt the breath of death on his neck in a more convincing manner than ever felt before.

The long-legged resident physician completed her examination. Murray's condition showed significant improvement from the prior day. His lungs were now clear and his edema and swelling of lower extremities had improved to a great extent. Murray had lost nearly 10 pounds of water weight during the night. Although uncomfortable, the Foley made it possible for Murray to sleep through the night. He was accustomed to an uncomfortable and frequent nocturnal urination and readily welcomed the rest. The systolic pulsation of the pulmonary artery was no longer palpable but his PMI was still quite displaced indicative of poor heart function.

Electrical History of the Heart
Murray's hospitalization was completed with a follow-up echocardiogram showing no further deterioration of his weakened heart. His ejection fraction was 25%. The electrocardiogram demonstrated a wide QRS duration and left bundle branch block indicating a very inefficient and lopsided beating of the heart. Murray's heart could be compared to a V-8 engine with two spark plug wires pulled off, causing it to run in a

rough and inefficient manner. The QRS duration on the electrocardiogram is preceded by the P wave. In Murray's case there was no P wave since he was in atrial fibrillation. There was no contribution of the upper collecting chamber filling the bottom chamber of the heart since the collecting chamber was now chaotic, quivering with inefficiency and clot potential. The width the QRS determines how long it would take for a contraction to complete a cycle. The longer it takes for a cycle to be completed, the less power the heart generates. The term for this is cardiac dys-synchrony.

In the 1700s, Luigi Galvani first verified that electricity was important in muscular function. He confirmed that placing an external electric current near the nerve of a muscle to a frog's leg would cause the leg to contract. Over 150 years would pass before Koehler and Mueller demonstrated that the heart was a producer of electricity. Their experiment placed a beating frog's heart next to another frog's nerve that supplied the muscles of the lower extremity. The heart generated an electrical pulse that caused the adjacent frog's leg to contract in a similar manner to an external application of electricity. An additional 30 years passed and the development of the capillary electrometer allowed Eintoven to record the first electrocardiogram and assign the names of the different waves of the electrocardiogram PQRST and occasionally U. These letters were assigned simply because Eintoven believed the beginning of the alphabet had been overused.

In the 1930s Dr. Wilson developed the 12 lead electrocardiograms, which are still in use today. This hundred-year-old technology is still quite useful in a diagnosis of arrhythmia and acute myocardial infarction and is one of the most misinterpreted tests in medicine. It is relatively simple, consisting of six waves including the U wave and 12 leads. The tracing can be used to diagnose hypertrophy of the right and left ventricles, atrial enlargement, congenital disease of the heart such as atrial septal defect, pulmonary embolus, and many other conditions. Interpretation of the electrocardiogram takes years of practice and exposure to many readings. Computers can provide fair but useless interpretations that many physicians tend to rely upon. Emergency room training programs teach electrocardiograms on the fly without formal teaching sessions and without great numbers of electrocardiograms to study. When the computer prints *Asterisks* * * * *, which usually implies a pending disaster; the reading doctor realizes the patient has a real problem. The interpretation of the electrocardiogram has fallen to reading a computer-generated diagnosis.

The cardiac conduction system, or the wiring diagram of the heart, consists of the sinus node which is a group of pacemaker cells that fire off more rapidly than the rest of the cells of the heart. Due to the speed of these cells, they become the commander and are in charge of the heart's contraction rate. Cells lower down in the system are depolarizing at a slower rate and can become the commander if the upper cells fail. Occasionally the cells become rebel cells and attempt to overtake command from the sinus node. Patients experience this with palpitations that can be quite troublesome to some individuals. This can also result in arrhythmia or a very fast heart rate. Generally when the upper cells fail to stimulate the heart, the rate will decrease because the cells are working at a slower rate. The sinus node sends several fibers out of the atria allowing the atrium to contract first as a booster pump. The electrical signal then travels through the atrial-ventricular node which is a pathway to slow the contraction. The slowing by the AV node allows the contraction of the atrium to finish filling the ventricle before the ventricle contracts sending blood to the lungs and body.

For the next 60 to 70 years this electromechanical coupling was for the most part ignored and only used as a diagnostic test. It was known in individuals with left bundle branch block, suffering a loss of part of their conduction system would have a greater chance of developing left heart failure. Children with complete heart block born from mothers with lupus erythematosus could also develop heart failure later in life. In patients who required pacemakers because they developed flaws in their conduction system; there was a greater tendency to develop heart failure than individuals without conduction system problems.

The DAVID trial came as a surprise to many cardiologists and to pacemaker companies. This trial examined patients with fancy defibrillators with dual chamber pacing at rates of 70 to 80 beats per minute as compared to patients who only had back up defibrillation with backup pacing at a rate of 40 or less. To the surprise of many, the more advanced pacemaker with continuous pacing did not demonstrate benefit. More people with this enhanced technology demonstrated more progressive heart failure and poor outcomes. The individuals who had intact conduction systems faired much better. These outcomes are explained by the cardiac dys-synchrony that occurs with right ventricular pacing. The group that received enhanced technology had right ventricular pacing. Right ventricular pacing is like removing 2 sparkplug wires from your V-8. The right ventricular apex is activated first causing a lopsided contraction pattern within the heart. This lowers the efficiency of the heart by the change in pattern of contraction and increases the duration of the contraction making it a less powerful pump. The solution to this problem is even higher technology adding a third lead to those individuals who require pacing. The patients who benefited the most from this technology are those who had left bundle branch block, and previous pacemaker dependent patients.

Murray had two electrical problems that would have to be addressed. The first was his atrial fibrillation and loss of the top-pumping chamber of his heart due to a chaotic electrical and mechanical movement. The second problem was his left bundle branch block causing cardiac dys-synchrony and worsening left ventricular function. After Murray had been placed on medicines and his heart failure optimized he would be assessed for the need for cardiac resynchronization and defibrillation to protect him from future sudden death and to help make his heart run more efficiently. Murray was pleasantly unaware of this plan. It would be addressed in the future allowing Murray to become use to the idea of having a constant companion in the pacemaker defibrillator.

Later in the day the cardiologist returned to Murray's room to discuss the treatment plans. The cardiologist wrote Murray's problems on a piece of paper. The problems included: 1) atrial fibrillation which is a fast irregular heart beat with an increased risk of stroke, 2) non-ischemic cardiomyopathy weak pumping function of the heart causing shortness of breath, edema, and fatigability, 3) cardiac dys-synchrony from his left bundle branch block, and 4) mitral regurgitation a leak of the valve because of the stretched dilated dyssynchronous heart. He was given a list of medicines with a titration scheme to gradually increase the dosage of these medicines. The slow deliberate titration of medications was important so that Murray's body could adjust to these changes. It takes time for the body to adapt to the changes and a rapid increase would cause side effects that would tempt Murray to throw the medicines into the toilet. It was explained to Murray that he would be followed in a congestive heart failure clinic by caring and attractive nurses who would ensure that he was taking his medications in a proper manner. He would also be receiving education in his diet, exercise, and that his smoking habit would be monitored. Later in his course of treatment there would be a plan to place a bi-ventricular defibrillator to restore cardiac synchrony and to act as insurance to shock Murray back to life if he should suffer a sudden cardiac death.

Philippine Challenge
The discussion then turned to the Philippines. Murray knew the cardiologist had been stationed in the Philippines. He had been to his office previously and noted the typical ornate nameplates that could have only been carved in the Philippines. He noticed the monkey wood staff the cardiologist used to discipline his fellows and the monkey head fertility hat that was used to help staff become pregnant and for good luck. The cardiologist, before he retired from the Air Force, served as the "Cardiologist of the Pacific" for the United States Air Force. He had been stationed with the 13th Air Force, Clark Air Force Base. The title of Cardiologist of the Pacific was a self-imposed title that meant he was the only military cardiologist in almost one third of the world. His patients came from Diego Garcia in the Persian Gulf, Korea, Japan, and Guam to just short of Hawaii. The type of patients the cardiologist would evaluate included both young and old airmen suffering acute myocardial infarctions before their time, native Filipina with the ravages of rheumatic heart disease, and neonates with complex congenital heart disease. The cardiologist was just out of training and had to live by his wits and nearby textbooks to manage such various conditions.

Murray asked the cardiologist, "Did you ever travel to Baguio in the Philippines?"

The cardiologist replied that he had been to Baguio. He traveled there as part of a consultation visit and to train individuals in advanced cardiac life support. Baguio the summer capital of the Philippines is high up in the mountains. Although the base of the mountain is hot and sweltering, the top of the mountain is cool with year-round spring breezes. The evening can become cold enough to support a fireplace. The city had both modern and primitive aspects. The large mansions displayed evidence of the great wealth and the small-overcrowded shacks portrayed the evidence for poverty. Silver was one of the mainstays of the population's craft and livelihood. Beautiful, delicate and intricate silver jewelry made of small chains was abundant in the marketplace. Woodcarving was also another occupation of the people. The art would range from very small nativity scenes to quite large animals and rocking horses carved out of a single piece of wood. Baguio had not achieved the modern city that is today at the time Murray had visited the area during World War II. The city in World War II had more of an ancient demeanor. Murray asked, "Would you like to go back and visit this area again?"

The Cardiologist sat back into his chair quietly for a few minutes and then responded "I would not enjoy the hordes of people in Manila, or the open air Jeepney's but I certainly would enjoy the picturesque landscape and the warm smiles of people. Yes, I would like to return. Murray then turned to the cardiologist and said if his heart could be repaired and he could make the journey he would like to invite the cardiologist and his residents on an expedition to the Philippines.

The discharge orders made an appointment in the congestive heart failure clinic within two days. The country boy in Murray made him want to escape to home and never return to the hospital. He really didn't want to take multiple pills every day. He didn't see how it was possible to carry these pills and still drive a tractor. Murray, however, had good remembrance of smothering on the day of his admission. It was the fear of smothering again that made him keep the appointment. The shiny black haired physician certainly made the hospital admission bearable. Her smile could even melt Murray's crusted heart. Her demeanor and confidence reminded him of his wife. She had been a female physician graduating in 1947 as an orthopedic surgeon. She had been confident and turned a blind cheek to any aversion or miss comment that lay ahead. She had been the perfect match for Murray to keep him in line. Since her death, Murray did not take care of himself nor did he manage his medication. Murray had to change his ways if he was to avoid future suffering.

Before starting Chapter 2, please note some comments from the author, Dr. Houck:

This book is written for my students, residents, fellows, and patients. Heart failure is a deadly disease but there is hope for the 5 million individuals with this disorder. Learn with Murray that medications, exercise, and attitude will extend life. Men, women, and children who wish to stay healthy will learn the benefits of prevention and exercise. This book is meant to be entertaining with education as a side affect. Hidden messages throughout these chapters will reflect that health care is an individual responsibility. The baby boomers are destined to destroy our country with a looming health care crisis, but it is not too late for the baby boomers to be a resource instead of a drain on society. Someone interested in the medical field may get a different sense for the practice of medicine. I hope some will change how they practice. I would be pleased if even one individual decided on a medical career. Mangers and policy makers may become enlightened to problems with no easy solution. The next President needs to tell the people what is needed and not just what the people want.

This book is for both medically inclined and the inquiring patients who want to know how the physical exam is accomplished and how the heart works. These sections are lectures that I have frequently given to my students and reflect the basics of cardiovascular medicine. Other chapters will reflect on medicine and the potential future of medicine. The glossary is for patients and lazy readers. Students should look up the answer and question any statement.

Frankly, I would like to change the world. The change has to occur in the individual, community and churches. Government provides infrastructure but solves few problems. I hope to change the reader so they stay healthy and productive beyond their retirement. "I" can't change the world but *we* can.

All of the characters in this book including the Houckster are an imaginary composite of my experiences. Placing a face to a disease process makes the diagnosis and treatment easier to learn. I am certainly not as confident, as good-looking, or as smart as the Houckster. As an author I have taken liberty to please myself.

While Murray begins his struggle to recover from heart failure, there will be a diversion to cardiology teaching rounds with Doctor Houck. The subject is dangerous, controversial, discriminating, and still poses great unanswered questions – Women.

Big Mistake

How do you treat her Heart?

Just Imagine she is a Man then treat her the same.

The morning rounds were completed at 11 o'clock and it was time for teaching rounds. Since the cardiology team consisted of a female senior resident, the rotating emergency room female resident, and one male intern, the cardiologist chose the topic of women and heart disease.

Colette, the emergency room physician had stepped into the examining room and had removed her jacket that was heavy with instruments, books, and notes. Her scrub top lay softly on her estrogen firm chest with the deep V-neck of her shirt revealing pleasant hills and valleys. In anticipation of his arrival, her breathing began to increase. The thought of the handsome aged cardiologist made her quiver. He had practiced what he preached to patients and had kept young by using weights 3 times per week, riding a bicycle, and swimming on a master's team. His shoulders were square and biceps stretched his medium size white coat. Although small in stature, his frame was large with broad shoulders and narrow waist. There was no six-pack, but the love handles were minimal. As he walked into the room Colette rose and rushed to greet the object of her desire. Just as her luxurious pink lips had found their mark she was startled awake with a choke, aspiration and a drool.

She heard the Houckster spit, "Minus 3 points for sleeping on rounds. You are now down to 6/10 on your continuous evaluation." stated the cardiologist. For years he had been evaluating residents and found it was much more effective to say minus 1 for simple errors and minus 5 for large errors. It was immediate and far more effective than yelling. Everyone started with a score of ten and felt lucky if they finished with positive numbers. Colette lost three points for sleeping and knew she would have to answer many questions that would test her arousal.

The cardiologist began with the following statement - Women are more difficult! Immediately the hackles of the three female physicians began to rise and a smile grew broadly across the male intern's face. Finally, he was hopeful, that he was going to get a break against the two beautiful and intelligent women. They were always successful in their central lines, intubations, and intravenous lines. They were always ready with answers during attending rounds. He felt that he could not compete against these intelligent and beautiful creatures. The cardiologist examined their startled looks and so he turned to the emergency room physician and asked if she felt male discrimination in the world of cardiology. The resident replied that she had been treated fairly and although there appeared to be a slight sex bias on some other services. She was unaware of any bias on her current service.

Discrimination
The cardiologist then stated, "Medicine and cardiology has discriminated against women." Often, discrimination is initiated with small acts, and no intended harm, and may not even be conscious. Discrimination in the health of women really was not a realization in the 1950s. It was not calculated but resulted from practicality with good intentions. In the 1950s it was common to see the breadwinner of the household die with an acute myocardial infarction. Women were unconcerned about their own health and were more worried they would lose their husband to the ravages of coronary artery disease. Myocardial infarction was serious business in the 1950s. The chance of survival prior to coronary care units and modern treatments was 50%. The high death rate during the 50's is compared to our current mortality in 2005 of slightly less than 5%. In the 1950s grandmothers would also have myocardial infarction and had even greater than 50% mortality, but because of their age it did not seem quite as tragic as losing a young breadwinner with three children who had college expenses. The goal then became to lower the death rate of the young male. Within the last half-century the myocardial mortality risk has been dropped by tenfold. The studies however, were performed predominantly in males. The results were applied to females with the hope they would respond in a similar fashion, but efficacy of these treatments were never tested in women. Women were not studied and this simple decision is the basis for our health discrimination against women. Other minorities are discriminated against simply because of small numbers and lack of enrollment into clinical trials. It seemed more tragic for the young head of a household to die leaving his family to fend for themselves than it was to say goodbye to Grandma. So for the next 30 years women were excluded from cardiovascular trials. The women were excluded from these trials because their inclusion would require the study to be much larger and much more expensive. The event rate for young women was quite small so including them within the study would be

a statistical nightmare. Physicians in the 1950's wanted to know what therapies would prevent myocardial infarctions in young men. At the same time they observed women dying of what they thought was an age related illness. The lessons that were learned in these cardiovascular trials were applied to women and minorities but outcomes to know if these therapies were appropriate were not studied.

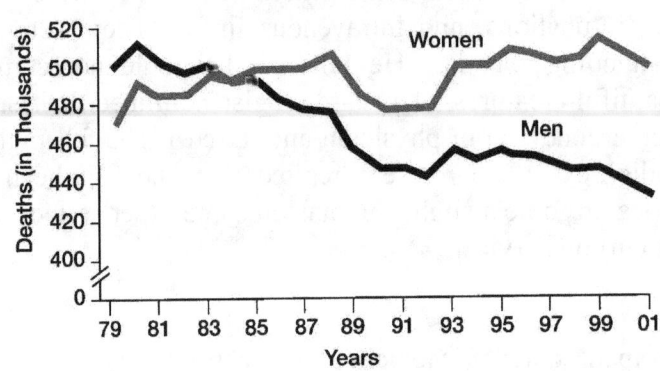

Deaths Due to Cardiovascular Diseases
United States 1979-2001

American Heart Association. *2004 Heart and Stroke Statistical Update*. Dallas, Texas AHA, 2003.

The end result of this bias is that in the last 20 years the mortality in heart disease in men has fallen as the mortality in women has climbed. Women were no longer content to pass on to the next world after they had lost their childbearing ability; they discovered they had a new life after menopause. Many times, women were taking over the care of corporations, grandchildren, or going back to school. They were no longer willing to simply just lie down and die with strokes and myocardial infarctions.

Attempts to correct this mistake have materialized in the last 10 years. A greater emphasis has been placed on studying heart disease in women. We have learned that treating cardiovascular disease in women is more difficult. A woman is more likely to die after myocardial infarction than a man. A woman has greater risk of dying during bypass surgery and angioplasty. Angioplasty is not as successful and has higher restenosis and adverse events. Women have more strokes. We have also learned that women show up to the emergency room one hour later than men following their myocardial infarctions despite having many more frequent visits to physician than their male counterparts. It may be due to 'tidy up the living room' before they allow EMS to enter the house, or that they are older and are lacking social structure to deal with emergencies. Older women are frequently alone and do not want to bother a family member when they are having trouble in the middle of the night. The female may be the caregiver of the family and feels that she cannot leave for her own emergency. Whatever the reason, females show up late and when dealing with myocardial infarction time is muscle loss. Minutes of delay can result in significant myocardial necrosis resulting in heart failure, ventricular fibrillation, and death.

Vascular Disease has Sex Differences

The vascular disease process is different in women than in men. According to the American Heart Association women tend to have more strokes and fewer myocardial infarcts. In year 2002 women had 372,000 strokes and 345,000 myocardial infarctions. In that same year men had 323,000 strokes and over 520,000 myocardial infarctions. Despite the vascular disease being similar in men and women there is a different presentation of the disease. Men suffer nearly twice as many myocardial infarctions than strokes whereas women have the same number of strokes as heart attacks. One third of women experience a cardiovascular event prior to the age of 65. Cardiovascular disease is the number one killer in women. 50% of women die from a cardiovascular disease condition. Twice as many women die of cardiovascular causes as compared to cancer deaths.

There may be a simple explanation for the difference in outcome between men and women. The explanation may be as simple as body size. Women are smaller and have smaller coronary arteries. A small river like the Guadalupe can be easily damned up as opposed to the Mississippi River. In the operating room, large coronary arteries are much easier to operate than a tiny small vessel. Inaccuracy in placement of a stitch is much better tolerated when the target is large as opposed to a very small blood vessel. One of the greatest advances in cardiovascular surgery was the development of microscopic glasses. If you walked into the operating room you would think the surgeons had very poor eyesight, because they're wearing eyeglasses which not only are simple glasses but have telescopic extensions to magnify the operating field allowing the surgeons to place a much more precise stitch. There are vessels that are so small they cannot be stitched without closing them. The angioplasty sheaths and balloons initially developed were too large for female blood vessels. The very first artificial heart was too big to be placed into a female's small chest cavity. Size does make an extremely important difference.

Estrogen - Good or Bad?

There is a female advantage. Examining every culture in the world today, women have an early advantage over men. Regardless of culture, dietary habits, poverty status, and activity level; there is a reduction of female deaths as compared to males early in life. The disease is not eliminated but presents approximately 10 years later in females over males and myocardial infarction lags nearly 20 years behind males. This female advantage is poorly understood. The result of this is that Grandma and her son could potentially die on the same date. Stem cell rejuvenation researchers have found that female donor progenitor cells repair tissues better than male donors and male stem cells are more inflammatory. The mystery of the female advantage continues.

In early retrospective trials it appeared that women who remained on estrogen therapy were protected from cardiovascular disease. Estrogen tended to lower cholesterol levels, elevate triglyceride levels and also increase inflammatory markers. Estrogen prevented hot flashes and may have given some degree of youth to menopausal women. Many of these trials were discounted since they were retrospective and it was felt that the females who took estrogen also had healthier lifestyles and were much more physically active. Their benefit therefore was not completely attributed to their use of estrogen. Most of these women began estrogen in the peri-menopausal time. The concept that women never went through estrogen withdrawal, has escaped the researchers who were trying to answer the question of estrogen. Is estrogen the reason young women do better?

The later clinical trials were performed in women who were "over the hill," breasts flopping, uterus shrunken and bulges in places where bulges should not be. Estrogen was reintroduced after a substantial time and it is likely that during the hiatus, estrogen receptors on many of the cells of the body disappeared. After menopause, large-scale trials were initiated to see if there is a benefit of estrogen in post-myocardial infarction patients. It seems odd to think that a medication that would increase inflammation and increase thrombosis would be considered beneficial in a post myocardial infarction female patient. These women had already suffered a thrombosis and likely already had a predisposition to thrombosis. In fact, these trials demonstrated a small very early harm from the estrogen therapy. As expected the increased inflammation and tendency toward clot formation raised the risk of thrombosis. There were more pulmonary embolisms and gall bladder attacks. By the end of the trial the harm seemed to have dissipated to some extent suggesting early harm with possible late benefit, but was sufficient enough to recommend that estrogen therapy in postmenopausal women after myocardial infarction is not beneficial. The study, however, did not answer the question as to why females have a benefit.

A follow-up trial of primary prevention, again, in postmenopausal "over the hill" women was investigated. A benefit might be present in these less thrombotic women. This trial was also negative and although important did not give us a clue as to why there is a female advantage. Post analysis of this same trial suggested that women who were closer to their menopause may have had an advantage from continued estrogen therapy. It does, however, point to the direction of future research. I consider the answer on estrogen still unknown, but estrogen is not helpful when you have already run "over the hill"

The answer likely involves both the key and the lock. Estrogen (the key) as well as the estrogen receptors (the lock) is important in mediating health and disease. Does continuing estrogen after menopause maintain the estrogen receptor? If estrogen and its receptors are equal (the right dose for the right number of receptors) will the female advantage continue beyond menopause?

With two major trials demonstrating possible harm from estrogen therapy, the demand for estrogen containing drugs plummeted putting many horses out of business. The estrogen had been most easily recovered from horse urine. With the decreased demand for horse urine, the horse no longer had a purpose, and became an expense rather than a profit margin. I am unaware of the fate of these horses - green pastures or cat food.

Some of the intuitive readers may wonder if estrogen will work in men. This was studied over 40 years ago and was found to be bad for men. Men do not have the receptors (Lock) for estrogen so they cannot benefit from the key.

Coronary Artery Disease in Men and Women
Coronary artery disease appears to be different in men and women. The subtle difference between men and women has been examined through autopsy evaluation of coronary arteries after death or with intravascular ultrasound (IVUS) during coronary angiogram. IVUS allows the cardiologist to look at blood vessels the same way as pathologist. Women tend to have more diffused atherosclerosis blockage in the blood vessels distributed at the end of the vessel where the vessel get smaller and smaller. The result is the proximal vessels may look normal on angiogram but blood flow is still impeded by blockages at the end of the vessel. Men tend to have more localized atherosclerosis in the proximal vessels. It is easier to bypass around this proximal stenosis than the diffuse disease that we see in women. Some women may go to cardiac catheterization and have normal appearing epicardial vessels but still demonstrate ischemia by scans and electrocardiograms. In the past these women were told that they were normal and not treated for their risk factors. Their prognosis was not as good as a normal coronary angiogram but not quite as bad as one demonstrating obstructive lesions. They were told the pain was not likely due to their heart. Today, we now realize that normal coronary arteries cannot be detected alone by simple angiography of the arteries. Intravascular ultrasound looks at the entire vessel wall where cardiac angiography only shows the lumen.

The etiology of myocardial infarction in younger women is different from that in older women. Younger women tend to have myocardial infarction due to plaque erosion. A plaque is a collection of cholesterol protruding into the lumen of the vessel wall. The artery responds by initially growing larger to maintain the lumen size. Eventually the plaque overcomes the artery and causes obstruction of flow. Plaque erosion occurs when the very fine single cellular layer of endothelium (the inner lining of the vessel) becomes denuded from the surface. Without the endothelium the vessel wall no longer keeps the blood from clotting. Platelets and blood clot will form on the eroded area and can cause a thrombosis. This process occurs in younger women and is frequently seen in young female smokers. In men and older women the process of plaque rupture occurs when the bubble of cholesterol explodes causing a crevice within this plaque exposing the contents. The thrombotic contents then cause a thrombosis. In an autopsy survey, approximately 25% of myocardial infarctions were attributed to plaque erosion. There were twice as many women than men with plaque erosion.

Women often have atypical symptoms that can be mistaken for other causes of illness such as:
•**Milder symptoms (without chest pain)**
•**Sudden onset of weakness, shortness of breath, fatigue, body aches, or overall feeling of illness (without chest pain)**
•**Unusual feeling or mild discomfort in the back, chest, arm, neck, or jaw (without chest pain)**

Based on the CASS (Coronary Artery Surgery Study) trial, when a male walks across the parking lot and experiences typical angina, stopping because of chest pressure and fullness, relieving the pain with nitroglycerin he will have a 93% chance at having coronary artery disease and a 7% chance of being normal with this classic symptom being a false positive test for heart disease. If a woman walks across the same parking lot and experiences the same typical angina, relieved with nitroglycerin, she will have a 72% chance of having coronary artery disease and 28% of the time will have a normal angiogram for this symptom complex. Women are more difficult to diagnose and therefore are often over treated and under treated based on their symptoms.

Heart disease in women is amplified by the coexistence of diabetes. Diabetes in women is much more deadly than diabetes in males. Over 4.2 million females have diabetes. This risk factor increases their chance of having a bad outcome 3 to 7 fold as opposed to men who have a 2 to 3 fold increase as compared to non-diabetics. 65% of all diabetics die of heart disease or stroke. Even women who had diabetes for less than four years had markedly increased risk for cardiovascular events. The Framingham study beginning in 1948 showed a relative risk of 5.4% in diabetic women as opposed to non-diabetic women. The nurse's health study showed a 6.3% increase in mortality in diabetic women as opposed to non-diabetic women.

Other differences between men and women include the risk associated with elevated triglycerides and low HDL. Triglyceride elevation and low HDL in women is much more deadly than it is in men. Women are also fatter than men with more than 25% considered obese and 52% considered overweight. There is a strong correlation with this excessive weight and the decreased physical activity recorded in the same women. Men were not much better with 18% being obese and 56% being overweight. Our culture may be guilty of sex bias. Females should be allowed to participate in lawn care. The distribution of physical activity and chores around the house should be distributed equally.

There is good news for women. The first Statin trial that included a large number of women was the AFCAPS trial (JAMA Vol. 279 No. 20, May 27, 1998) with 997 women and 5,608 men. The women showed an impressive 46% drop in events when they were taking Mevacor. Men also had an impressive improvement at a 37% drop. Women appear to respond to Statin therapy in a very vigorous and highly impressive manner. Perhaps their small blood vessels that are easily clogged can be unclogged with greater efficiency. Other drug trials in women may have raised more questions. Ace-inhibitors, beta-blockers, IIb IIIa inhibitors, and Digoxin did not seem to give women the same benefit as men. In most of these trials one of the reasons may be the inadequate number of women who were tested.

Medications Perform Differently in Men and Women
A large trial in women was the Women's Health Study (Ridker PM, Cook NR, Lee IM, et al. A randomized trial of low-dose aspirin in the primary prevention of cardiovascular disease in women. *N Engl J Med.* 2005;352:1293-1304.). The results of this study further emphasize the need to test medications in both women and men. The results of this study shook the medical community, because the results were unexpected. One of

the oldest medications we currently use today is aspirin therapy. If aspirin was brought before the Food and Drug Administration today, it is unlikely to be approved because of its numerous effects and would not be considered for over-the-counter use because of its tendency to cause gastrointestinal hemorrhage. It is however a life-saving medication and has been shown to prevent re-infarction and primary infarction in men. Many individuals have improved their outcome from myocardial infarction by taking and chewing an aspirin at the onset of their chest pain. The purpose of the women's study was to evaluate similar findings in women. The study was to be performed similar to the Physicians Health Study that had investigated aspirin.

Bayer invented aspirin in 1897. The Physician's Health Study was in predominantly males. This study was performed between 1982 and 1988. The Women's Health Study was not initiated until 1993 and concluded in 2004. There were differences in how aspirin was administered. In the men's study a full 325 mg tablet every other day was used. In the women's study a smaller dose of 100 mg of aspirin was used every other day. The smaller dose was known to inhibit platelets but was hoped to decrease the amount of gastrointestinal bleeding. Critics claim the study may not have used the correct dose of aspirin. The study was also criticized to have included too many healthy women in the study. The studies although slightly different were designed for safety. The results were very surprising demonstrating unexpected lack of benefit in females. Under the age of 65 no benefit from taking aspirin as primary prevention was noted in women. However, in women over the age of 65 aspirin was proven beneficial.

The poster child for the woman at risk for coronary artery disease is a fat, diabetic, hypertensive female sitting on her butt smoking.

Unfortunately, women are not the only group that has been discriminated against in the treatment of heart disease. Due to a lack of participation, or, recruitment in clinical trials minorities have not been studied. Again, the intention was not to discriminate but to facilitate. Therapies are applied to all races, but proof of benefit exists mainly for white males. Race specific studies had not been performed until recently.

In the study of congestive heart failure etiology there is a difference between the self reported black (African American) and white (Caucasian) populations. Self-reported black is the term used in the A-Heft trial to describe their study population. Black and white will be used for this discussion. The white population has predominantly epicardial coronary artery disease as the major reason for cardiac disease. In the black population, there is less epicardial disease but more hypertensive heart disease. The high incidence of hypertension in the black population is conjectured to be a result of a survival advantage in retaining salt and water. The slave galley's had high death rate due to inadequate supplies of water. The survivors had the ability to retain salt and water. In a society with great salt excess this advantage has turned to a disadvantage due to the resultant hypertension.

The etiology of congestive heart failure in these two populations is therefore, quite different. In the white population, beta-blocker therapy and ace inhibitors have been shown, without a doubt, to cause an increase in survival among patients suffering from congestive heart failure. The very first study in heart failure compared vasodilator therapy with Hydralazine (anti-oxidant and vasodilator) and nitrates (Nitric oxide donor) versus Prazosin (an arterial vasodilator) versus placebo. In V-HeFT 1 trial, ace inhibitors were pinned against the then standard therapy of Hydralazine and nitrates and were an overwhelming winner. Nearly 20 years later the study was reanalyzed and it was found that the entire benefit from the ace inhibitors was seen in the white population. Re-analysis of the first vaso-dilator trial demonstrated that most of the benefit was seen in the black population. This prompted the A-HeFT trial be performed in black Americans. This trial demonstrated that the combination of hydralazine and nitrates given three times daily was very effective in preventing death, adverse events, and hospitalizations in the black population. The drug tested was Bidil, which is a combination of nitrate and Hydralazine. It has been FDA approved for race specific treatment of congestive heart failure. However, it should really be approved as a disease specific treatment of congestive heart failure, heart failure secondary to hypertensive disease and not as a race specific disease.

According to recent genetic mapping, we all have a little bit of African heritage since this is the likely origin of the human species. White patients with hypertensive heart disease should be treated with Hydralazine nitrate combination and Black patients with epicardial coronary disease and systolic dysfunction should get ACE inhibitors.

Question Your Teachers!
Thus the main focus of this chapter is we all fall into traps, and even attempts at improving health can take unfortunate detours. We have realized our mistakes and are taking steps to correct injustice to women and minorities. Diligence in the science of medicine is a never-ending battle. Question your teachers! Question what you have been taught. Leave the world a better place.

Discharge Home for Murray The Work of Healing now Begins
Murray had been dried out with generous use of Lasix and was off his oxygen. He received his discharge instructions and his appointments. He would be entered into the heart failure clinic where he would receive titration of his medications and more dietary reminders to stay away from salt. He would be followed in the anticoagulation clinic to monitor his Warfarin dosage and in 6 weeks to two months be cardioverted back into sinus rhythm to help restore cardiac efficiency by allowing his atrium to stretch his left ventricle prior to its contraction. In three months he would receive a bi-ventricular pacemaker/defibrillator to restore cardiac synchrony and protect him from sudden cardiac death. The next three months belonged to the health care system. Murray would be performing cardiac rehabilitation and at least weekly visits to the health care system.

The next three months would be difficult. Improvement of fatigue and shortness of breath would come very slowly. Some days would be worse than others. He would forget to take his Coreg with food and get dizzy weak and light headed. He would feel weak every time his medications were changed or when he took too much Lasix. After Thanksgiving and a wonderful salted feast he would be edematous and short of breath on the following Saturday. No pain no gain.

Murray was overwhelmed by the number of appointments and the time he as going to spend in the clinics. Part of him wanted to give up and die like an old soldier. Part of him remembered the drowning sensation he had when he was admitted by the black haired beauty. Part of him wanted to have one last adventure in the far-east recapturing the thrill of adventure and the unknown. Murray was now determined to improve and he would obey the medical orders like the crusty old soldier that still lurked inside his weakened body.

Chapter 3
Coronary Artery Disease

Myocardial Infarction Number 1

Beep. Beep. Beep. The MI pager sounded off in a sobering manner. One telephone call had the ability to bring seven people to the same single location - the cardiac catheterization laboratory. The system was designed to decrease the time it takes to treat myocardial infarction. Time is muscle. Time is muscle! A longer delay in opening up a blood vessel results in more myocardial cell death and worse outcome. The first to arrive at Emergency Room Number 2 was the cardiology fellow. The cardiology fellow was popular among the staff. He was in his third year of training and had studied the heart night and day for 3years. He was pleasant to the staff and very decisive in his demeanor. Being 6 foot 2, with dark complexion he was voted most likely to become the centerfold of Playgirl Magazine by his internal medicine residency class. His looks and demeanor helped him obtain cooperation among the staff. To become a cardiologist, the fellow had to attend 12 years of normal education, four years of college, four years of medical school, three years of residency, and, three to four years of cardiology fellowship. The sum total is 27 years of training without a paycheck for 21 years and a pitifully small check during residency and fellowship. All this training with very little pay increases indebtedness. Moonlighting becomes necessary to pay the bills.

The Emergency Room doctor flashed the electrocardiogram towards the fellow. The ST elevation in the inferior leads could be recognized across the room. The diagnosis was straightforward and treatment was non-controversial. The electrocardiogram is a technology that is 100 years old and is also the most likely reason for a mistake to be made in emergency room. The biggest EKG mistake is not repeating the electrocardiogram when pain continues. The second biggest mistake is to misinterpret the segments that can be quite subtle in their changes. The difference between an expert reader and a novice is the prior electrocardiogram. The old electrocardiogram allows comparison so that event subtle changes can be obvious.

During the initial presentation to the emergency room, even with the patient in pain, the electrocardiogram may be normal or non-diagnostic. Changes of a myocardial infarction take time to evolve. There are also some heart attacks that can occur with normal electrocardiograms. These elusive, difficult to diagnose heart attacks can be detected by suspicion and a good echocardiogram. An echocardiogram is produced by a $200,000 portable machine that can take ultrasound pictures of the moving heart. This tool can detect abnormal wall motion. A more time consuming method of confirming heart muscle injury is to draw blood cardiac enzymes that rise to abnormal levels indicating that heart muscle has died. Obtaining cardiac enzymes requires blood draws six hours apart to give time for the enzymes to become elevated in the blood. A good cardiologist never waits for enzymes to become positive because time is muscle. As the blood vessel continues to close, the electrocardiogram becomes quite clear. However, this can only be identified if the electrocardiogram is obtained.

The fellow had completed his task of making the correct diagnosis and performing a physical examination within two minutes. The next task would be much more difficult and involved obtaining consents from the very uncomfortable patient. He would simultaneously order the appropriate anti-coagulation with Heparin, Aspirin, Plavix and Integrilin and check if the cardiac catheterization crew had been notified to come to the cath lab for this emergency.

Informed consent is a discussion between the physician and the patient about the risks and benefits of the procedure that is planned. Many patients, however, lack the capability of understanding their disease process and rely on the judgment of the physician. The legal system demands that this consent process be undertaken and documented. At times, the patient is too ill to give consent so the family member accompanying the patient, who is also under duress, grants the consent. Only lawyers concern themselves with the appropriateness of these signatures and only if the outcome is not good. The patient should understand the most important issue is that they could die from the procedure but their chance of dying without the procedure is greater. Other harm can occur in rare instances and is usually the result of bad luck. Such instances include significant problems such as stroke, renal failure and possible allergy to the contrast media. Within five minutes the cardiology fellow reassured the family and obtained the consents for angiography (the process of taking pictures), angioplasty (the procedure to fix the blood vessel), and blood transfusions if surgery was necessary or bleeding occurred.

The fellow ordered the appropriate amount of Heparin, Integrilin, Aspirin, beta-blocker, and Plavix. He checked the renal function to be sure the doses were appropriate. These drugs would help combat the thrombosis that was ongoing in the patient's coronary artery. Thinners are used to prevent clot formation. The small molecules IIb IIIa inhibitors are to block the function of the platelets that are the initiators of the heart attack. Platelets are the tiny elements smaller than blood cells that circulate through the blood vessels. Their job is to look for any disruptions in the endothelium, the single celled lining of the blood vessels. If there's a disruption, such as an injury, the platelets swarm to this position and help dam the blood vessel to prevent blood loss. Later, the proteins in the blood responsible for clotting form strands to further enforce the dam constructed by the platelets. Platelets are beneficial to prevent excessive bleeding, but are disastrous in the heart when they attach to abnormal endothelium over an atherosclerotic plaque resulting in occlusions on the unhealthy endothelium. The endothelium is the largest organ of the body and is responsible for our growth, development, and supplies nutrients to all cells of the body. These cells are long-lasting and live nearly 25 years before being replaced by new young cells. The endothelium is remarkable in that it is the only substance that can come in contact with blood without forming a clot. The endothelium is also the source of stem cells that help repair our aging organs.

Joe's Initial Presentation
The patient, Joe, was a 55 year-old truck driver who was a smoker, obese, hypertensive, and ate only fast foods on the highway. The amount of salt in his diet was evident by the salt crystals lodged in his beard. His family history includes his father suffering a myocardial infarction at age 50. His father died aged 56. One of his brothers underwent

coronary bypass surgery in the recent past. He had been driving down the highway when he noted the rather sudden onset of the deep sickness within his chest that felt like a pressure radiating up into his throat and into his left arm. The drenching sweat, accompanied by a feeling of doom caused his clothing to be wringing wet and cool. The lightheaded and woozy feeling made him feel like he was leaving his body. He had radioed for EMS and pulled off the side of the road.

Within 10 minutes, two minutes slower than recommended in the guidelines, the EMS had located his truck on the side of the road and began to transport him to the emergency room. He arrived in the emergency room within 15 minutes and had a total duration of 30 minutes of pain. His electrocardiogram was obtained within five minutes of reaching the emergency room and the cardiac catheterization laboratory was notified one minute later. Another five minutes later the patient's anticoagulation therapy had been initiated and by 15 minutes his pain began to resolve as the thrombus in the blood vessel began to weaken and allow fresh blood to travel down the coronary artery.

When the blood vessel opens up dramatic things can occur. An open blood vessel helps restore blood flow to the starved myocardium. The danger occurs when the blood vessel opens and reperfusion arrhythmias occur. The reperfusion arrhythmias are ventricular tachycardia and ventricular fibrillation. Ventricular tachycardia is an organized fast heart beat at 150 beats per minute that may cause the patient to pass out or feel very weak. Ventricular fibrillation is a chaotic rhythm that always results in pass outs or death. If these rhythms are not treated quickly, death will ensue in 10 minutes. These rhythms are popularly displayed on hospital programs. The patients will be shocked and jump high into the air making the gurney shake. In most instances a single shock will restore the heart to normal rhythm. Any delay in performing the shock will result in metabolic deterioration that makes the rhythm more difficult to terminate.

The truck driver received textbook treatment. He was unlike most males who would have ignored his symptoms and likely have wrecked his truck on the highway due to sudden cardiac death. Males are more likely to present with sudden cardiac death than females. The presenting symptom of coronary disease as an irreversible event such as death or myocardial infarction occurs 62% of the time in males and only 40% of the time in females. It seems that females may have a better sense about impending doom and seek help. Their demise is that they always seek help one hour later than the males who do decide to go to the emergency room for their chest pain complaints. Because time is muscle, females have a worse outcome from their myocardial infarction including death and congestive heart failure.

The hour was late, 0100 in the morning. Despite this inconvenient time, the cardiac catheterization crew had arrived to transport the patient to the catheterization laboratory and had 15 minutes of the golden hour to be sure the blood vessel was open. As they were wheeling the patient towards the catheterization laboratory, the staff doctor met him in the hallway and did a brief examination. It was after this examination he observed the female patient in room 1.

Myocardial Infarction Number Two Alice

She was an elderly 78 year old female appearing to be in distress. He recognized her as being a former hospital employee who retired many years ago. Taking a few moments to say hello, he inquired as to why she was in emergency room. She stated that she had sickness. She had not felt well the last three weeks, had increasing fatigue and had given up her daily two-mile walk. Earlier in the evening she developed lower chest and abdominal discomfort with sweating and nausea. She experienced similar feelings while walking during the previous three weeks. She failed to share this with the emergency room doctor because she did not think it relevant. She traveled to the emergency room by private automobile and was triaged to abdominal pain where she waited for 6 hours. The cardiologist asked the emergency room nurse to perform an electrocardiogram on this patient. The EKG demonstrated extensive anterior ST segment elevation indicating a very large myocardial infarction. Since time is muscle, a large portion of myocardium was already been lost. The new guidelines have changed what patients are supposed to do. For years it was recommended that nitroglycerine should be taken every five minutes then go to the Emergency Room. Now, a single Nitroglycerine is taken and EMS called. The change was to get patients to the Emergency Room quicker, safer, and triaged sooner than in the past. Alice, although a seasoned medical personnel, did everything wrong.

Women and men often experience the same symptoms in the presentation of myocardial infarction. Symptoms may include:
· **squeezing, or stabbing pain in the chest**
•**Pain pressure or fullness, radiating to neck, shoulder, back, arm, or jaw**
•**Pounding heart, change in rhythm**
•**Difficulty breathing, shortness of breath**
•**Heartburn, nausea, vomiting, abdominal pain**
•**Cold sweats or clammy skin**
•**Dizziness**

Women often have atypical symptoms that can be mistaken for other causes of illness such as:
•**Milder symptoms (without chest pain)**
•**Sudden onset of weakness, shortness of breath, fatigue, body aches, or overall feeling of illness (without chest pain)**
•**Unusual feeling or mild discomfort in the back, chest, arm, neck, or jaw (without chest pain)**

Alice was an example that women are more difficult in diagnosing coronary artery disease and acute myocardial infarction. The night had now become more complicated.

The race for resources, and triage had now evolved into a difficult decision process. There were now two individuals, one an elderly female near the end of her life, and a young male both requiring the cardiac catheterization laboratory. The staff doctor stopped the gurney and inquired to the truck driver how he was feeling. The truck driver indicated that his pain had vanished. The staff doctor asked for a repeat electrocardiogram that demonstrated resolution of the ST segments. He then informed

Joe, the truck driver, that his immediate problem had showed improvement and there was more desperate a patient that also required cardiac catheterization laboratory at this time. Joe understood the situation and agreed to allow Alice, the retired nurse to go first. "Doc, I feel great, just do what you have to do."

In the meantime, the cardiology fellow initiated the required consent process and anticoagulation therapy. He quickly reviewed the existing laboratory and history. Alice's gurney was unlocked rolled toward the Lab. Alice did not receive guideline therapy. She made the mistake of coming to the Emergency Room on her own. She delayed activation of EMS. Her abdominal complaints resulted in her being triaged to low priority instead of high priority. An electrocardiogram was not performed in a timely manner because of the atypical presentation of a myocardial infarction. Six hours had been wasted by not establishing a rapid diagnosis. The patient also delayed an additional two hours because of her hesitancy to seek attention. Alice, a trained and intelligent female was transformed into an ostrich with her head stuck in the ground.

Three times per week, before she retired at the age of 65, she and her husband would dance for two hours at a time. She retired to care for her husband of 45 years when his Alzheimer's disease became too advanced for him to remain at home alone. Prior to retirement, she and her husband made an attractive and graceful couple on the dance floor. She would have continued her nursing career but now was nursing her husband who asked the same questions multiple times per day. One of the reasons Alice delayed her visit to the emergency room was to get her daughter to come sit with her husband. Alice gained 30 pounds in the last five years due to a lack of exercise. She knew better than to add extra salt because of her borderline hypertension but could not resist the addictive crystal. As with many health professionals, she failed to obtain her annual exams and was not screened for her elevated cholesterol and elevated inflammatory markers that could have identified her as an individual with increased risk. Preventive medicines were not part of Alice's nature. Her family history was significant with her older sister having a myocardial infarction three years ago. Alice completely understood and gave a little smile when the Houckster informed her she had picked the wrong sister.

Weather and Myocardial Infarctions
At one o'clock in the morning the cardiac catheterization crew did not enjoy having their work doubled. They looked at the old cardiologist with evil eyes, who responded with a kind smile and gleaming eyes, stating there will be a third case before they could go home. The crew wanted to know where the patient was coming from and the cardiologist replied he had not received the call, but felt the *force* (clairvoyance) and was quite sure the patient was on his way to the emergency room. The attractive, blond, scrub nurse with bed hair sticking out in odd attractive poses and no chance to shower, asked the cardiologist in a sweet Texas drawl, "Did this have to do with the low-pressure front that had come through yesterday?" She was aware of the publication **(Relation of Atmospheric Pressure Changes and the Occurrences of Acute Myocardial Infarction and Stroke.** The American Journal of Cardiology, Volume 96, Issue 1, Pages 45-51 P. Houck, J. Lethen, M. Riggs, D. Gantt, G. Dehmer) brainchild of the Houckster that related the increased frequency of myocardial infarction to a drop in atmospheric pressure. The greater the drop in atmospheric pressure, the greater the number of

myocardial infarctions expected the next day. With blue eyes fluttering, she asked the cardiologist how anyone thinks up such a lame brain idea. After all, you can't see atmospheric pressure and nobody can do anything about the weather anyway.

While the patient was being prepped for the diagnostic angiogram, the cardiologist began his tutorial in the development of original ideas. Original ideas are often like a lightning bolt striking from heaven. You never are quite sure where they come from but you know they are accurate when the bolt hits the ground. Background and training have a great deal to do with new insights. If one is only trained in a single discipline, individuals tend to think merrily within this discipline and often reject ideas outside of this area. It takes a broad background, such as that of the Houckster who initially trained at Penn State University as an engineer learning mechanical forces and later at Northwestern University in Biomedical Engineering to think out of the box.

While I was at Northwestern University in Evanston, Illinois, I realized a bioengineer could do very little without the help and direction of a physician. This realization eventually directed me to medical school and further training in cardiology and study of the heart. The heart is a marvellous pump and requires both the knowledge of electrical forces, mechanical forces, fluid dynamics, and biology to completely appreciate this marvellous organ.

The power of atmospheric pressure came from personal experience. My father was tank platoon commander in World War II and lost his right leg. The discussion of his Silver Star and Purple Heart never occurred until I returned from college. My father never spoke of the trials and tribulations he had to endure in his youth and only looked to the future. I remember as a child getting angry with my father's teasing and popping him on the right leg and finding to my great surprise that the leg was as hard as wood and invincible, making my father impervious to the pain. This lesson was followed with great respect for my father and never again did I attempt to pop him on his right leg. My father had been asleep in a farmhouse in Germany with the rest of his tank crew when a bomb or shell hit the farmhouse (probably friendly fire according to my father) with a piece of shrapnel detaching his right leg, with numerous other shell fragments that would work their way through his body to the surface over the next 50 years. The hot pieces of shrapnel seared the vessels saving him from hemorrhage. He was airlifted back to the states and eventually to Walter Reed Hospital. My father received life-saving penicillin that had been reconstituted through recovery of the drug from other soldier's urine. He spent many months of recovery and adjusting to his newly disabled state. He lost weight from 160 pounds to just barely over 100 pounds. His strong will and faith allowed for his recovery then he entered the workforce as a bank teller and later in life became a pillar in society with many people seeking his wisdom. During his life, however, he spent many sleepless nights; he related the insomnia to phantom pain due to changes in the weather. The phantom pains, like searing hot pokers, untouchable itches, relentless tingling, and paresthesias would come from his leg and robbed him of precious sleep. I was impressed that my father could predict the weather and the weather changes could make him so uncomfortable. I learned respect for the weather and the pain it can cause. Other weather related disease processes include individuals with sinus pain; precipitation of labor and childbirth, increasing suicide rates. These ailments and many others could all be correlated to the unseen changes in the weather.

Alice gave a brief "yelp" of pain as the numbing needle slid into the skin. Unconcerned, the Houckster continued his lecture as the fellow continued to numb the skin in preparation for the Cook Needle to penetrate the femoral artery. At Penn State, in one of my first mechanical engineering courses studying static forces I learned that bathyspheres were more likely to fail upon ascent than descent. I was impressed because I was aware of submarine movies and crush depth that they always passed to make the movie more exciting. It now makes the movies even more exciting when I realized that as they passed the crush depth that they had false sense of security. They made it past that depth but still may never see the surface again. One of the greatest engineering feats that gave the Russians an edge was their ability to dive to greater deaths by their ability to cut holes, portals into their submarines. At any place within the submarine where there is a crack or a hole, this is where the most likely place that failure occurs. For these reasons the portals are reinforced. The coronary artery is no different from the submarine in its response to pressure loading. The origin of failures would occur in plaques that had thinned to 65 μ or less or had defects within the plaque making them more susceptible to failure. These defects include the vaso vasorum, which are tiny blood vessels that travel through the atherosclerotic plaque. Other defects include neovascularization (tiny new blood vessels) that are associated with inflammation and even crystallization of cholesterol into sharp edges under the influence of decreased pressure.

For another 25 years my knowledge of my father's leg pains and engineering of bathyspheres would be dormant, and eventually, come together to form a hypothesis for acute myocardial infarction and the weather. The necessity for forming some type of hypothesis became relevant early in my career as a cardiologist. Patients would ask a simple question, "Why did I get a heart attack?" For many years the only erudite explanation was bad luck. Patients would accept this explanation but it certainly was not a satisfying explanation to someone who was intent on studying the disease for the rest of his life. I would explain to victims of myocardial infarction that inside their blood vessel was a bubble of cholesterol surrounded by a fibrous cap and for some reason the fibrous cap would become thin and rupture. This rupture would cause a thrombosis within the artery. The villain in the blood stream was not the blockage in the artery but the reason for the blockage to rupture and the subsequent clot. Numerous studies document observations but do not explain why the fibrous cap ruptures. The question needing answered was the reason for the fibrous cap to rupture. There have been many observations that have been reported in the literature and confirmed in multiple studies. Patients have more heart attacks in the morning and thus demonstrate a diurnal variation. This means waking up can be hazardous. Hospitals are also busier in the winter with more frequent myocardial infarctions. It does not matter if winter is in Maine or in Hawaii where the temperatures were quite different. Snow shoveling tended to be associated with more myocardial infarctions even though the method of snow removal made no difference, by a machine, a shovel, or the neighborhood boy next-door. It was not shoveling, but the snowstorm that was the culprit.

Alice thanked Dr. Houck for putting her to sleep with his ramblings during the painful part of the procedure but claimed she would appreciate a gas next time instead of hot air. Unperturbed by Alice's comments, the Houckster continued as the guide wire and the

Left Judkins Catheter slid up the aorta from the femoral artery dodging cholesterol laden mountains in route to the left main coronary artery.

Careful observation has always been a route to discovery. My observation came during sleepless nights when thunderstorms would pass over and awaken me from a sound sleep. After numerous years of caring for the many patients in the coronary care unit, I began to expect the next day to be eventful with multiple patients arriving in the emergency room at the same time with myocardial infarctions. In our own hospital, the average number of myocardial infarctions per day was between one and two. After a strong weather front often accompanied with thunderstorms, it would not be unusual to admit 5 patients with a heart attack within a brief amount of time. After the passage of harsh weather conditions it is not uncommon to see two people needing the cardiac catheterization laboratory at the same time. I formed a hypothesis that the fall in atmospheric pressure associated with the low pressure front, oftentimes a thunderstorm, would cause the atherosclerotic coronary plaque to rupture similar to the bathysphere. The time from the rupture of a plaque until myocardial infarction becomes clinically evident is unknown. Based on my nocturnal observations, I could estimate that after the plaque ruptures it would take approximately 12 hours until chest pain and myocardial infarction became clinically evident.

I decided to study my hypothesis by obtaining atmospheric hourly pressure recordings for three consecutive years and correlating this to the number of myocardial infarctions and strokes that occurred over the same three-year time. The study was limited because the precise time of myocardial infarction was not recorded, only the day of the event. The amount of data was too large to be included in one Excel sheet. The data sat for nearly 5 years until I was introduced to a post graduate student at Texas A&M University who was looking for a statistics project. The result was a published paper demonstrating the weather affects on myocardial infarction and a completed masters degree.

The next research step would be a bold interventional attempt to decrease the number of myocardial infarctions. The intervention would be selecting patients at risk and pre-treating them during times of rapid falling atmospheric pressure with aspirin, Plavix or both. The success of this study depends on correctly identifying patients who would benefit from this therapy. In this potential study, the top 30% of the highest risk individuals for myocardial infarction are chosen from the population, researchers will treat them with aspirin/Plavix in hopes of saving more lives by prevention of myocardial infarction then causing deaths from gastrointestinal bleeding. The therapy is worthwhile if the number of lives saved is much greater than the lives lost. Treating the entire population including the low risk population in this manner could potentially cause more harm than benefit.

Technology Gap
One major drawback in the current health care system is that patients are unaware of their own cardiovascular risk. There is a technology gap at the level of the patient. Patients who have high risk are often unaware. Some who know their family history and their risks are like Alice, an ostrich with its head stuck in the ground. The health-care system tends to only treat people after the patients become ill; like this poor young lady on which we were performing primary angioplasty.

Cardiac Catheterization

In the last 10 minutes the crew had been re-educated why they were in the cath lab at 1 A.M. The sleepless night was the fault of the weatherman and not the old grey haired cardiologist. During those 10 minutes, the patient had been scrubbed, instrumented with monitoring leads, and draped with a sterile blue paper. The right groin had been pricked with a tiny needle that allowed the burning local anaesthetic to blunt the pain of insertion of the introducer, a large IV placed into the femoral artery. The arterial wall cannot be adequately anaesthetized so when the Cook Needle enters the vicinity of the artery the patient usually has a small reaction- "OUCH!" After the brief initial pain the procedure is painless since there are no nerves inside the blood vessels. An experienced hand can quickly feel the pulse and guide the needle toward the red spurting fountain. A wire is placed into the artery to act as a guide to place the introducer and catheters.

With the first injection of the left main through the JL4 Judkins catheter, it became obvious the old nurse had a "widow maker" or in her case, a "widower maker"- proximal obstruction of the left anterior descending. The right coronary was studied and was without significant blockage. In fact, the only vessel diseased was the left anterior descending. The job of the cognitive cardiologist was now over and the interventional cardiologist took over to complete the procedure. The fellow and the interventionalist worked quickly getting a wire down the occluded vessel followed by a pre-dilating balloon. With the vessel opened, oxygenated blood and free radicals poured into the occluded segment with electrical irritability. A few PVC's (extra premature rebel beats) followed by a sustained run of ventricular tachycardia causing the old nurse to lose blood pressure and consciousness. The technician in the control room was already on her feet and in 15 seconds 150 joules of biphasic energy was delivered to the chest and apex with a successful shock. The blood pressure remained at 40 mmHg. 4 units of Vasopressin restored BP to 100/60. The stent was deployed without further problems and the vessel was patent but had stuttering flow down the vessel with persistent dye staining of the myocardium. Both of theses signs indicate the tissue being perfused was swollen and likely not to recover. Alice, the old nurse began to cough and have increased respirations. The sick ventricle was no longer emptying the blood and pressure in the left atrium was building spilling fluid into the lungs. The distress of the pulmonary edema caused the blood pressure to climb. A serious and poor prognostic sign, pulmonary edema (a rapid suffocation with internal fluids) was ensuing. Alice did not want to be intubated. She had seen too many individuals suffer on a ventilating machine.

Lucky for Alice, Natrecor had been invented. This is a polypeptide produced by the heart and consists of a hundred peptides. Brain Naturetic Peptide (BNP), the formal name of the hormone, attempts to protect the heart. It would cause both venous and arterial vasodilatation. It also would oppose the evil neurohormones and would act as a mild diuretic to help remove excess fluid. It was a coronary dilator and a bronchodilator. The drug was administered IV and she almost immediately began to breathe slower and better. In 20 minutes she had died of ventricular fibrillation and had been shocked back to life, nearly drowning in her own body fluids until Natrecor was administered, and was now feeling pretty good. Alice was feeling better with the help of the Natrecor and she thanked Dr. Houck and the fellow for the good care and the education. She was transported to the CCU so the truck driver could now take her place.

Cardiac catheterization has truly evolved since Dr. Sones at the Cleveland Clinic had a medical mis-adventure. Dr Sones was performing a ventriculogram with an end-hole catheter when the catheter jumped out of the ventricle and dove into the right coronary artery filling the vessel with the powered injection. Dr. Sones stopped what he was doing, put down his cigarette, and waited for the anticipated ventricular fibrillation. This would be a death event since defibrillator machines had not yet been invented. To his amazement nothing happened. He inadvertently performed the first coronary angiogram. The first angioplasty for myocardial infarction was in a patient who was admitted with unstable angina and his diagnostic angiogram showed a 95% right coronary artery obstruction. He was scheduled for an angioplasty. One hour before the angioplasty he had chest pain and ST elevation. Instead of canceling the procedure, he was brought to the cath lab and successfully performed a balloon procedure. Medicine and good luck sometimes travel together.

Joe's Turn in the Cath Lab back to Myocardial Infarction 1
The cardiac lab turnover was efficient, even for 2 o'clock in the morning. Joe, the truck driver who was waiting in the emergency room was quickly prepped and the procedure initiated. The right coronary artery remained open and thus did not suffer any further chest pain. Benefit was probably obtained by the cooling off time delay so the anticoagulant medications could help reduce the clot allowing less debris to embolize down the vessel. Small particle emboli occlude the small vessels causing obstruction and swelling in the microcirculation. Placing a stent across the ruptured plaque, results in a strainer like effect sending broken clot and debris from between the struts to the distal circulation. The strained debris clogs the microvasculature causing infarction (multiple small infarcts). This particular area of cardiology has always been controversial. If the patient has stabilized, is it better to treat the patient medically for several hours or days? Some cardiologists withhold the invasive procedure altogether. My particular preference is to treat the infarct vessel early to prevent re-infarction. Most of the morbidity and mortality occurs when the vessel re-infarcts. A re-infarction occurs when the vessel closes down after the initial opening. Since we can't predict to whom and when this will occur, I tend to take most of the patients to the lab early. There is a plethora of questions needing to be answered. One question addresses the open vessel hypothesis. An open blood vessel promotes healing, less arrhythmia, and LV function may be better preserved than if the blood vessel is left closed. The advantage of a closed blood vessel is the vessel cannot close again with a catastrophic infarction. The Open Artery Trial (OAT) trial has recently given some insight to this complex problem. This trial attempted to answer this question in ST elevation myocardial infarction when the vessel was closed for greater than 48 hours. In this select group of stable patients with a closed artery leaving the vessel closed was better.

Myocardial Infarction Number 3 Jethro Q vs Non-Q
As the cardiac catheterization laboratory crew was preparing to go home at 2:30 in the morning, the cardiology fellow's beeper again raised the adrenaline with the screeching "beep, beep, and beep". Remembering the old cardiologist prediction, a collective sigh of dismay filled the room and no one would dare to look at the staff cardiologist sitting quietly in the corner. A frantic intern on the fifth floor of the hospital made the page. A patient had been admitted earlier in the day with multiple episodes of chest pain during the previous three days redeveloped chest pain at two o'clock in the morning. The pain was an uncomfortable tightness associated with shortness of breath. The pain was worse

lying flat (supine) and much better with sitting up. The patient was sitting up in bed gasping for breath and would not lie down. The pain and uncomfortable feeling improved, but was not relieved with nitroglycerin and morphine. The patient's blood pressure was now 90 systolic. An electrocardiogram demonstrating worsened ST T-wave abnormalities changed from the admission electrocardiogram. The pain had been relentless for the last 30 minutes and his chalky lips tainted by Maalox revealed that reflux was not the source of his pain.

The patient was a 67-year-old recently retired schoolteacher, Jethro, who had a long history of indigestion and gastroesophageal reflux disease (GERD). His current symptoms appeared similar to his gastroesophageal reflux. Initially his electrocardiogram was normal with only subtle changes, and his cardiac enzymes were negative. According to the admitting physician, the symptoms were somewhat atypical for cardiac pain. The chest burning and heaviness would occur soon after he would recline to bed, and was not a great deal of bother when he was physically active. His physical activity during the recent weeks was quite limited and consisted of getting up from his chair to go to the bathroom. This recent decline in activity was due to a skiing accident and a fractured ankle. The schoolteacher's admission diagnosis was Acute Coronary Syndrome, with a high suspicion that his symptoms were due to gastroesophageal reflux disease (indigestion from acid spilling into the feeding tube). For this reason, the admitting doctors decided not to treat him with anticoagulants or anti-platelet medicines while they awaited further cardiac enzymes. His disease process was no longer a mystery and his diagnosis was now was changed to a non-Q wave myocardial infarction because of the unrelenting pain and EKG changes. Q wave myocardial infarctions tend to involve the inferior wall with ST segment elevation in the inferior leads and occasionally posterior ST segment depression. Anterior myocardial infarctions involving the left anterior descending or diagonals cause ST segment elevation in the precordial leads. Non-Q wave myocardial infarctions may involve the circumflex, which is relatively silent on the electrocardiogram. Non-Q wave infarctions can also occur in both an inferior and anterior distribution. In these two distributions, the infarct is threatened and not completed. Instead of the vessel being entirely closed off by thrombus, the ruptured plaque simply serves as a source of emboli to the distal vessels causing pain and EKG changes that are different from dramatic ST changes in Q – wave infarcts.

Cardiology and cardiac disease have evolved greatly over the last century. Thirty years ago, the most prevalent myocardial infarction was ST segment elevation Q-wave myocardial infarctions. The reason for this change in disease process is unknown, understudied, and is likely due to multiple reasons. Prior to the advent of cardiac enzymes many non-Q wave myocardial infarctions simply were never diagnosed. Other reasons for the change in prevalence include the greater use of aspirin, and earlier recognition of symptoms by patients. Patients are now presenting earlier in the infarction process so that the infarct did not have the time to completely close the vessel.

The staff doctor who heard the conversation of Fellow had already arrived on the floor before the flushed, sleepy intern had gotten off the phone. Just as the intern completed the call he was standing face-to-face with the staff doctor. The intern commented that the service by staff doctors was really outstanding in this hospital and wondered did he ever go home? He also knew that even though was to 3:00 the morning he would have some education forced down his throat.

The staff doctor asked the intern on call what kind of heart attack he would like to have, a ST segment elevation infarct (Q-wave) or a non ST segment elevation infarct (non-Q wave)? The intern had been under previous inquisition from the staff doctor and was a little intimidated to answer. After a certain measured silence the intern replied he would rather have a non-Q wave myocardial infarction since the ST segment elevation infarct seemed much more dramatic and required thrombolytics or immediate transfer to a hospital with a cath lab. The staff doctor replied that he was correct but only for the first 24 hours. Only in the first twenty-four hours death was more likely in a Q wave infarct as opposed to a non-Q wave. However, at six months more people with non-Q wave myocardial infarctions would be dead than with ST segment elevation infarcts. So the answer at end of six months was a Q wave myocardial infarction. The intern was trapped again.

The staff doctor and the intern walked to the patient's bedside. The fellow soon arrived. The staff doctor asked a few simple questions while gazing at the fresh electrocardiogram. He then pulled three consents out of his pocket and handed them to the fellow to complete the consenting process. The staff doctor informed the patient, that the best course of action was to proceed with immediate cardiac catheterization to open his closed blood vessel. He excused himself, to ensure the cardiac catheterization crew had not escaped during the brief lull in activities. The efficient catheterization crew had already anticipated that they would still be busy and had already turned the room around for the next anticipated case.

On the floor, the patient received a bolus of heparin with a drip, Integrilin bolus and drip was started and the pharmacy orders checked to ensure aspirin was administered. Anticipating bad anatomy, the patient had not been given Plavix. The electrocardiogram demonstrated multiple leads showing abnormal ST statements including ST elevation in lead AVR suggesting a large area of ischemia or multi-vessel coronary artery disease. Mr. Jethro Hendrix was transported to the cardiac catheterization laboratory prepped and draped in the usual manner. With the initial injection, there was a soft *"Oh, shit."* barely audible in the deathly silent room. The initial injection demonstrated a 90% distal left main occlusion. The collective deep sigh of all the personnel in the room unfortunately gave a strong signal to the patient that all was not well. One other orthogonal picture was taken of the left system demonstrating distal target vessels to be of good calibre. The right coronary artery was non-dominant small vessel, making the left main even more critical. The patient had been within minutes of dying from the left main occlusion. Jethro had a disease process that was very serious, but undoubtedly, gave the best survival benefit from surgical procedure. There was no doubt the coronary bypass surgery would be prolonging his life. He was lucky he was still breathing.

An intra-aortic balloon pump was placed. The balloon is a 40 cm balloon that inflates in a timed sequence in the descending aorta (main artery from the heart) and pushes blood towards the head and the heart while the heart is resting during diastole (the time the heart fills as opposed to the time the heart is contracting (systole). This improves the blood flow to dangerously narrowed coronary arteries. It also makes it easier for the heart to pump. When the balloon deflates, the heart sees 40 cc of empty space and it is easy to fill empty space as opposed to a vessel full of blood. With initiation of the intra-aortic balloon counter pulsations Mr. Hendrix's pain resolved. After a brief discussion with the cardiac surgeon, it was decided the patient had been stabilized and would be sent to the coronary care unit to wait for surgery the next morning after the Integrilin infusion had been discontinued for two hours. Although the operation could have been performed immediately, the outcome produces better results if the surgeon is rested. The time is now 4:30 a.m. Instead of going home, the crew decided to shower at the facility and get ready for the next day's work. Working all night and all day is not a desired element yet is part of the territory. The staff doctor thanked the crew and said he would try not to call them until the sun rose. They should also remember *"Red in the Morning Doctor's Warning."*

Atherosclerosis – 75% of Us have the Disease
After a brief shower and oatmeal from the cafeteria chased by a headache-curing tall coffee, the staff doctor felt mildly refreshed and ready to start the day. Since we have been busy throughout the night with three myocardial infarctions, inferior, anterior, and non-Q myocardial infarct we should spend some time discussing coronary artery disease and its natural history. The Houckster turned to the raven eyed emergency room resident and asked her "When does atherosclerosis (build up of cholesterol blockage) begin?" In a luxurious and low pitch voice, the resident replied that myocardial infarction's occurred earlier in men than women and that she guessed that the atherosclerotic process began in the fourth decade. She felt she was currently immune from coronary disease, but was quite concerned about her grey-headed staff. The Houckster replied that coronary atherosclerosis might begin before you are born. The first risk factor for coronary artery disease is low birth weight or pre-maturity. Individuals tend to have abnormal metabolism that follows them the rest of their life after this unfortunate early life experience. The disease has been documented in very early stages of fetal life. If you were able to inspect the blood vessels of the fetus of a mom who has severe hyperlipidemia (cholesterol), fatty streaks could be observed in the blood vessels within the first three months of the life of the fetus. The bad news is coronary atherosclerosis begins early. The good news is that by the third trimester as the fetus matures its own metabolic pathways and the fetal lipid metabolism can cause regression of the atherosclerotic fatty streaks. Atherosclerosis can disappear.

This knowledge is not new. The knowledge that atherosclerosis is present in young individuals has been present for more than 50 years. In the Korean War, soldiers who died went to autopsy and the pathologist found coronary atherosclerosis and blockages in 75% of those young males aged 23 years of age. Unfortunately, soldiers are always young.

The Houckster paused and reflected on his childhood to make a point, "I remember going into my parent's attic and looking at old *Look* or *Life* magazines that had a picture of a soldier staring into space. The story was about the early atherosclerosis seen in our soldiers and the explanation was the stress of war. The American public could not accept the fact that they all had the disease. There had to be a reason these young soldiers were afflicted. The finding of early coronary disease was attributed to the stress of war. The study was again conducted during the Vietnam War and this time post-mortem angiograms were performed. The angiogram pictures of the coronary tree again showed 45% of the soldiers had atherosclerotic blockage. Follow-up autopsies demonstrated the same 75% incidence. There was a discrepancy between what the pathologist could find and the angiogram. The pathologist can always find more atherosclerosis than the cardiologist because the cardiologist in the early days could only look at the lumen (tunnel) of the blood vessel. As the blood vessels become abnormal with atherosclerosis, the blood vessel remodels in a positive way becoming larger preserving the lumen to blood flow. Since cardiologists don't like to have anyone do anything better than they can, intravascular ultrasound was developed to look inside the blood vessels to give the cardiologist the same view is the pathologist.

In our modern world Ackerman examined young vehicular accident victims, victims of urban warfare who died in automobile crashes. Again, 77% of these young patients had coronary blockages. Nissen of the Cleveland Clinic examined teenage heart donors and 1 out of 6 had coronary blockage.

If we examine the entire population in the United States, three-fourths of the individuals will have coronary blockages. This is roughly 175 million people. The number of heart attacks and strokes per year, however, is smaller than that prevalence of disease. In the year 2002 there were 1.5 million strokes and heart attacks, affecting only 2% of the at risk population. Besides atherosclerosis, cholesterol blocking of blood vessels, there must be another key to this puzzle. The key to the puzzle is the mystery that causes the atherosclerotic plaque to rupture or to erode with resultant myocardial infarctions and strokes. In recent observations, patients suffering from myocardial infarction and stroke have had inflammation of the blood vessel which is suggested as the cause for the weakening of the plaque. Inflammation of the blood vessels indicated by a blood test known as High Sensitivity CRP is a very strong predictor and even a stronger predictor than Cholesterol/HDL ratio.

The study of heart disease has evolved from the time of Heberden's first description of angina:

**"There is a disorder of the breast marked
with strong and peculiar symptoms
considerable for the kind of danger
belonging to it, and not extremely rare,
of which I do not recollect any mention
among medical authors. The seat of it,
and the sense of strangling and anxiety,
which it is attended, may make it not
improperly be called angina pectoris.
Those, who are afflicted with it, are
seized while they are walking and
most particularly when they walk soon
after eating, with a painful and most
disagreeable sensation in the breast,
which seems as if it would take their**

**life away, if it were to increase or to
continue; the moment they stand still,
all this uneasiness vanishes."**

Heberden, W. *Some accounts of a disorder of the breast* Med Trans Coll Physicians
(London). 1772; 2:59-62

Risk Factors

The Framingham study, initiated in the late 1940s, was the first attempt to get a handle on the cause of atherosclerosis and the reason for heart attacks, strokes, claudication (leg pains with walking), and other diseases of the great vessels. The study involved slightly more than 5,000 inhabitants in a small village of Framingham, Massachusetts. Approximately one third of individuals refused to participate in this study, and these individuals tended to have a greater higher or increased would fit better death rate than those individuals who were studied. The term "risk factor" was coined from this study. This study is ongoing, now including the offspring of the original group of participants. As the most frequently cited study, it has provided the medical community with great insights into the possible reasons for the atherosclerotic process. The study was limited in the early days of investigation due to the lack of sophisticated tests. Stored blood samples have been used and newer testing performed on some of these highly studied individuals. The number of risk factors suspected for coronary disease was quite limited. The study of lipids was not fully developed. The findings of this study define risk factors by correlating lifestyle habits, blood pressure, and laboratory factors with events occurring within this population. Simple parameters were determined: age, gender, cholesterol, blood pressure, diabetes, and smoking. Age is obvious; old people die. Age is not considered a modifiable risk factor. Picking one's parents is very important in avoiding heart disease; however, family history is not modifiable. Gender is not modifiable. Although sex change operations may change gender, the risk for vascular disease reverts to the gender at birth and not the post-operative result. (The last statement is a supposition and not backed by any clinical study.)

Modifiable risk factors include cholesterol, blood pressure, diabetes, smoking, obesity and exercise status. The public who will be future patients are not aware of their modifiable risk factors. They do not know their blood pressure, cholesterol, and their only clue to diabetes is frequent urination. The current patients who have been told their risk factors forget what they were told and do not modify their smoking, diet, and exercise lifestyle. The medical community, perhaps from economic, social, or a rushed practice, under treats many of these patients failing to get them to published goals. Only when the public assumes responsibility for their individual health and the medical community assumes responsibility for prevention of disease will the epidemic of vascular disease decline.

New risk factors are being discovered and the links to cardiovascular disease are slowly being delineated. These are known as emerging risk factors. The population as a whole has changed with the percentage of obesity increasing. The increase in per person body weight will eventually translate to more diabetes, more heart disease, and a fall in the life expectancy. This projected fall in life expectancy has never happened in recorded history.

Homocysteine, a by-product of metabolism, was initially raised as a risk factor in the 1950s. This was at the same time that the time the cholesterol model was proposed. Atherosclerotic plaque could be produced by feeding rabbits a diet high in Homocysteine. The debate in the literature allowed cholesterol to win, and for 25 years the Homocysteine model lay dormant. Cholesterol does not explain all coronary disease. It is likely there are other inherited metabolic problems that will explain causes of infarctions in small populations. Other emerging risk factors include: markers of inflammation, clotting factors, and special sub fractions of lipids like lipoprotein (a).

The well-studied risk factors from Framingham cannot explain all vascular disease events. Most heart attacks occur when cholesterol is within the normal range. Our purported normal values have also come from this study. Recently, these normal values have come into question. Are the individuals who lived in Framingham, Massachusetts and did not have a heart attack or stroke healthy? Or were they just lucky? These individuals reflect the culture, diet, and exercise status of the population during that time. What LDL (bad) cholesterol is considered normal? The guideline recommendation for desired LDL cholesterol in patients with disease or with diabetes has continued to drop. For those people without risk factors, LDL of 130 is considered desirable. The LDL of 130 is only an assumption and is based on the lucky (healthy) individuals in Framingham who did not have a heart attack. The LDL you were born with was 50 to 75. Even this low LDL is compatible with the exponential growth. Indigenous populations, although shrinking in size, living off the land, calorie poor have LDL of 50 to 75. Therefore, it should be concluded that the people in Framingham, Massachusetts who did not suffer a myocardial infarction death should not be considered normal only individuals who had a slightly less chance of having an event. The normal LDL should be 50 to 75.

History of Cardiac Catheterization

The treatment of coronary disease over the last half-century sprang from daring deeds and miss-adventures. In 1929, Werner Forssmann was fired after he and his unsuspecting nurse passed a urethral catheter through a cut down in his brachial vein into his right heart. He walked up stairs and had an X-ray taken to document the course of the catheter into his heart. He demonstrated that pressures could be measured in the heart without killing the patient or himself. His fellowship director fired him for this act but later re-instated him.

In 1958, Dr. Mason Sones, invented coronary angiography with a mistake. Now that blockage in the arteries could be identified treatments would follow. On the backs of Alex Carrel, Vineberg, Sabiston, and Bailey; bypass surgery using saphenous vein anastomosis from the aorta to the distal coronary artery was successfully reported by Favalaro and Johnson in 1968. Bypasses initially were performed off pump because the Heart-Lung Machine had not yet been invented. The heart lung machine allowed greater care in placing the tiny stitches into the millimeter sized arteries. History has a means of repeating itself and the old technique of off pump surgery was re-invented with some technological improvements. The new techniques are now performed with robots and octopus clamps through small incisions. After surgeons learn to do open procedures, cardiologists attempt to follow with closed procedures. Andreas Gruentzig, a German physician, developed a technique of dilating arteries with a balloon attached to a small catheter. 30 years ago in Zurich, Switzerland he treated the first male with this technique and that individual is still alive today. Both bypass and angioplasty have evolved and are the most common operation performed today.

Genetics of Vascular Disease
The understanding of coronary artery disease continues to evolve and is becoming interdisciplinary. The Human Genome Project has opened up new avenues in the discovery of acute myocardial infarction. Within some families, acute myocardial infarction appears to be an inherited legacy. Parents, brothers and sisters are oftentimes afflicted by this disease at a young age with few family members living beyond the age of 60. Family histories that are so common for cardiovascular tragedy have been a source for study and perfect for gene analysis.

Dr. Eric Topol previously at the Cleveland clinic has performed extensive research on a family from Iowa that has tragic history. By examining their DNA structure searching for similarities and dissimilarities between other patients' groups, he was able to identify a myocardial infarction gene hidden among the twenty thousand genes that determine our nature. The defective gene on chromosome 15 is called MEF2A and is involved in the repair of arteries. This gene was discovered in a myocardial infarction ridden Iowan family. It may account for up to 2% of heart disease in the general population.

Other mechanisms for myocardial infarction are being discovered. A gene that may affect nearly 10% of individuals was discovered in Iceland. This gene was found by looking at the genetic material of individuals in Iceland who had myocardial infarctions. This approach was more of a gunshot approach that yielded a gene that is more likely to affect more patients. This gene acts in an accelerated fashion and increases inflammation within blood vessels. The mechanism blocking the hyperactive response of the gene is currently being investigated and may pave the way for direct therapy to individuals who inherit the wrong genes.

Two of the genes now identified have implicated inflammation and vascular repair. Due to the complexity of these processes it is likely many other genes exist that may explain why only 2% per year of the affected population has myocardial infarctions. As these mechanisms are delineated there is hope that myocardial infarction could be prevented. The hidden secrets of our DNA molecules are slowly being revealed. The end result may be that a few individuals will have the knowledge that they can be fat without exercise and not have risk of myocardial infarction because they were lucky to not have received the wrong genes.

While gene therapy has promise, it has not delivered any therapies to the clinical world. An abnormal gene either fails to produce or over produces molecules that are used in normal cellular metabolism. An individual is born with this abnormal gene that fails to transmit information or transmits miss-information within the cell. The result is a production of proteins or changes in a receptor within the cell wall. The production of proteins or the change in receptors results in the disease process. One method of treating this problem is to repair the gene abnormality. The repair would have to be in every cell of the body. How to change the genetic code in every cell of the body has been the great barrier to gene therapy. Evolution may be a form of gene therapy. By mutations and viral infections inserting information into the genetic code, new proteins and new functions developed. If the mutated individual had a survival benefit their lineage would continue and be part of our current genome. Humans have attempted to use viral infection with imbedded code to affect a cure. It has not worked – yet. (**Survival of the Sickest**: A Medical Maverick Discovers Why We Need Disease, by Dr. Sharon Moalem, Jonathan Prince, a Hardcover from William Morrow) is a fun and excellent resource.

Understanding the genetic process and how proteins are made can help find treatments of disease. If a protein is not produced and results in disease process; potentially a medication can be given that would supply this needed protein. Other abnormal genes may change receptors and their function. Identifying abnormal genes will lead to personalized medication.

The adverse responses to medications experienced by some individuals can not be avoided. Currently the only way to know if someone is going to have an adverse reaction is to give the medicine and stand back waiting for the complaints to materialize. By knowing who those individuals are before exposing them to the drug will individualize drug prescription.

In this chapter we have learned that coronary disease is a result of genetic predisposition amplified by risk factors. Nearly 75% of us have coronary disease - blockage. While there are no clear explanations as to why some individuals have a heart attack, the process likely involves inflammation and mechanical disruption of the fibrous cap that covers a cholesterol pool, and perhaps vulnerable blood that is predisposed to clot. Treatment is best served by avoiding a heart attack by having a healthy lifestyle and certainly by not smoking. Since Mason Sones accidentally injected the right coronary artery, cardiologists have been treating blockage. The blockage, however, may not be the primary evil villain in this disease. The villain still to be determined will likely be better treated with medication then bypass or angioplasty.

Today, angioplasty and bypass surgery have a life saving role in acute myocardial infraction and acute coronary syndromes. In stable angina they can only relieve symptoms and not prolong life. Extensive coronary disease with heart dysfunction has mortality benefit from coronary bypass surgery. More diffuse disease as in diabetics also benefits from bypass surgery. Angioplasty and stenting relieves symptoms and in some cases of less severe disease is as good as surgery. In the future, with exercise, diet, good lifestyle, and medications to correct genetic deficiencies, the barbaric splitting the chest and bypass grafting may become of historical interest relegated to the Smithsonian Institute.

Murray is now ready to encounter the heart failure clinic where he will be educated in the medications that are necessary for his recovery. Hatred and fear of medications is almost universal. Everyone is suspicious of Drug company motives and wants to have free pills but medications are the way to relieve suffering when there is illness.

Chapter 4
Medications

Adriaen Brouwer in The Bitter Medicine(1635)

Tracy stumbled into the heart failure clinic at 7:30 in the morning. Her curly hair had so much body that her head seemed disproportionately small. Her eyes were a luxurious brown-green color with the complexion that could only be described as vanilla ice cream; pretty to see and even better to taste. She was well conditioned due to her jogging with long muscular shapely legs. She had a determination and drive that allowed her to work full-time, study for Ph.D. level coursework, and still raise a family. She was a single mom with two overachieving children. Exercise is one of the things that kept her sane with her busy schedule.

She had only four hours of sleep the night before, because she was finishing a paper for one of her Ph.D. courses.

Her nurse training included early exposure to a country nurse who would have felt comfortable wearing a paper hat. Older nurses tended to be trained with quite harsh discipline. They were not only loving, caring nurses; but, had to be prim and polished while performing quite disagreeable tasks. She trained Tracy early in her career and left impressions that Tracy would hold onto the rest of her life. Nurses are trained to be patient advocates. Even among nurses there are different breeds. Some nurses follow orders dutifully but are uninterested in the complex physiology that is happening to their patients. Other nurses have curiosity to understand the course of therapy. Tracy possessed great curiosity. She had worked with the staff cardiologist on many occasions in the coronary care unit. At times, Tracey was concerned about the novel therapy he administered. She would question him as to the reason for using a particular medication. A short physiology lesson would ensue and she would gain understanding from it. From these multiple brief encounters she learned to trust his judgment and his ability to pull patients back from the brink of disaster.

It was pure luck that she and the Houckster came together to initiate the congestive heart failure program. He had the talent to design medication schemes for the patient's and she possessed the organizational talent to put his design into action. She would pool resources from the entire allied health services. The proper care of congestive heart failure patients includes the ability to educate the patients in lifestyle changes. She developed handouts and visual aids to show the salt content in common foods. These were displayed in a diagram with salt containers that expressed salt content in a dramatic physical manner and not just as a misunderstood number. Her friendly and caring demeanor allowed the patients to trust her immediately.

The patients who listen to her benefit from her knowledge and have fewer episodes of heart failure. Her job also included titrating medications for patients. In order to ensure proper adjustments, she had to confirm patients had the means to obtain the medications. She spent countless unpaid hours working with social work and drug manufacturers to obtain free samples and free medication for patients who could not afford these life-saving medications. Some patients can afford medication but feel their money should not be utilized to purchase something as mundane as medications. She would even help this population obtain samples understanding their particular personality quirks. She didn't care that she may have been duped; she only cared if patients would take their medicines and benefit from taking them.

She relied on allied medical resources including cardiac rehabilitation, dietary services, social services, physical therapy, occupational therapy, and home health to investigate home situations. Working with patients tends to be a frustrating experience. The obstacles to better health include simple lack of intelligence, personality disorders that displace all responsibility to others, social economic issues that prevent obtaining medications, and lack of simple bathroom scales (or scales that do not go over 350 pounds), or scales that can not be used because the patient can not stand. The process of caring for heart failure patients involves many aspects of their life. She put the cardiologist's medication protocols to practice in a practical manner individualizing to the patient's quirks. The cardiologist's job was easy. Putting his plan into successful action required a great deal of work that Tracy loved. She would see patients on a frequent basis meeting weekly at first, and then monthly until their medications had been titrated to the appropriate levels. If patients were not improving as expected she would send them back to the cardiologist to look for a reason for their failure. The philosophy was to approach the patients as a team with the only goal of improving the patients' lives. This team approach would allow 30% of the heart failure patients to have their heart function improved entirely with dramatic improvements in the ejection fraction from 30% to the normal value of 60%. This improvement usually occurred six to seven months after therapy is initiated and titrated to the full extent. Further therapies were initiated including biventricular pacing and defibrillators. Another 30 - 40 % of patients improved to an asymptomatic status but still had degrees of heart dysfunction. The last 30% of patients tended to have further deterioration frequent hospitalizations and ultimately die of their condition. These 30% were always the most challenging patients and required even more tender loving care to help improve their condition. These patients could sense that they were in dire straights and would want to know the number of days they still had on Earth. Prognosis in this group is exceedingly difficult to assess because some patients with ejection fractions that are very low at 15 to 20% would have continued longevity with no further deterioration. Other patients, for no apparent reason, would have a sudden decline and miserable existence. Heart failure patients are unpredictable with unexpected success and tearful failures.

Patients are given the medicines to be titrated over the next three months. Tracy listens to their complaints tenderly, reassures the patients and encourages them to take medications even if they feel the medications make them feel poorly. Tracey maintained the patience necessary to understand and relate to patients. Tracey's multifaceted personality combined this understanding with cunning diplomatic tactics and occasional temper tantrums to produce desired behaviors from the patients. Tracy is one of God's Angels placed on earth to take care of the rest of us.

Murray blundered into the small examination room. He sat in a chair and looked at the blank computer screen adorning the desk and the examination table to his front. He saw the blood pressure cuff and had a transient feeling that he should get up and run out the door. As his agitation and anxiety were increasing, the door squeaked open and Murray felt his day change for the better. Tracy walked in with a smile that was endearing over perfectly enameled teeth. She walked to Murray; clasped his hand and said, "Welcome to the heart failure clinic. May I have your cup please?"

Murray handed her the 32-ounce Slurpie he had purchased from 7-11 that morning. In a very casual manner Tracy walked to the sink and poured the contents down the drain. Smiling broadly, she turned towards Murray. "Did they forget to give you your discharge instructions?"

Murray would have stood up and placed a bayonet into her heart if she hadn't been so cute and attractive. Murray was old and considered over the hill but still felt excitement in the presence of a feisty young female. Tracy made her point with a smile instead of frown and a shaking finger, thus winning Murray's confidence. He quickly fell in line and apologized for his oversight. He was just having trouble believing all the salt and water nonsense.

Tracy began her typical introduction to the congestive heart failure clinic. She told Murray that an educated patient is more likely to take their medications and is more likely to survive their illness. The prognosis in untreated heart failure is similar to cancer. The medications we use will help you feel better, but it takes time for the medication to work. Some of the medications will not make you feel well initially but will slowly improve your heart so you will live longer and better. If you cannot afford the medication, we will help you obtain them from drug company programs.

Diuretics

The initial form of therapy for congestive heart failure is to relieve congestion with a diuretic (water pill) called Furosemide, more commonly known as Lasix. This medication allows the kidneys to excrete excess salt and water. Diuretics are critical in the improvement of symptoms including: shortness of breath, inability to lie down, and leg swelling. This medication, however, has no mortality benefit and can cause harm over long-term use. Sodium, magnesium, potassium, and other salts are lost in urine. Patients can then become dehydrated and develop renal dysfunction because the kidneys are starved of blood flow. When a patient is dehydrated from Lasix they will feel short of breath due to a ventilation perfusion miss-match. Lasix is a medicine heart failure patients always need to keep nearby because of possible dietary indiscretions that occur especially around holiday celebrations where excess salt is consumed. The heart failure patient has a difficult time excreting salt, thus resulting in an accumulation of salt within all body fluids over time. The salt will hide in all liquid components of the body including inside cells. The distribution is everywhere so it is difficult to detect. In order to maintain electrolyte balance, the body swells with fluid to keep its electrolyte levels normal. Lasix (Furosemide) is a diuretic typically prescribed for short-term use. Unfortunately, it is commonly continued past the time of its beneficial affects. If the patient feels dizzy or if leg swelling goes away the patient should stop taking Furosemide. If the patient is compliant with giving up salt, Furosemide will no longer be required.

Lasix (Furosemide) is best taken while in the supine position (flat in bed). In this position, the kidney receives greater blood flow and will work more efficiently to eliminate excess salt and water. Lasix is to be taken early in the morning. The patient rises, relieves the bladder, and returns to bed. The alarm clock of the next full bladder will determine when you can get out of bed.

For even greater urinary elimination of salt and water, Zaroxyln is provided 30 minutes prior to taking Lasix. Zaroxyln affects the proximal tubules of the kidney and provides more sodium for elimination at the distal tubules where Lasix works. This medicine is necessary only every other day or every third day for patients who remain noncompliant with their salt use.

Another diuretic occasionally used is Diamox. This medication can be used once per week and should not be used daily unless closely monitored. Diamox allows for the excretion of bicarbonate into the urine. With volume depletion, bicarbonate levels can rise resulting in decreased breathing patterns. This process is known as respiratory compensation. These patients typically have other problems such as obstructive sleep apnea from obesity or central hypoventilation. A majority of congestive heart failure patients may have sleep disordered breathing. (Javaheri S, Parker TJ, Wexler L, et al. Occult sleep-disordered breathing in stable congestive heart failure. Ann Intern Med 1995;122:487-492.) The fatigue of heart failure is partly due to inadequate sleep.
Obstructive sleep apnea patients and obese patients may be placed on a diuretic for peripheral edema. These groups can have a paradoxical affect from the diuretic because

the edema is not a vascular fluid overload, but a tissue fluid overload. Edema can occur from obstructive sleep apnea or venous insufficiency when the valves in the veins deteriorate and allow the long column of blood to exert great pressure on the capillaries in the lower extremities. Central obesity will elevate the venous pressure. As a result, the pressure forces the fluid to leak into the tissues. The patient can therefore have total body fluid overload with edema but inadequate fluid in the blood vessels. When diuretics are given for these conditions it results in hypertension. The kidney sees less blood flow because of the diuretic and its response is to raise the blood pressure. While diuretics are typically the mainstay in the treatment of hypertension, it is important to note that in some subgroups of patients, diuretics may result in hypertension. Diuretics given for venous insufficiency and edema not related to congestive heart failure can be very harmful to patients. Patients often demand a diuretic, "I am not fat, I am retaining water and I need a strong fluid pill." They are correct that they are retaining fluid but it is because they are obese.

Joseph G., Joe for short, worked as a short order cook at a neighborhood restaurant. The restaurant could hold almost 20 individuals at one time but was generally packed during breakfast, with a moderate lunch business. The evening meals were usually local retirees who decided it was easier to come to Joe's than to cook a meal on their own. Reasonable prices allow the clientele to eat at his establishment and still pay for some of their medications. To help make ends meet, Joe would buy less than prime cuts and obtain older discounted vegetables that he could enhance with the age-old culinary secret of salt. By adding salt to most of his dishes there were very few complaints about the inadequacy of the taste.

Unfortunately, Joe would eat his own food. As result of the salt loading, he developed hypertension at the age of 40 and was untreated for the next 15 years. During those 15 years Joe's heart became thick and stiff from the hypertension. Joe's leg swelling progressed to weeping and he was having difficulty putting on shoes without becoming significantly short of breath. He eventually went to his physician and was prescribed Furosemide as a diuretic to relieve the excess of salt. Joe never returned to his physician, since the Furosemide was quite efficient at removing the excess salt and fluid. One morning, Joe woke feeling weak as he struggled to get out of bed. He was woozy and lightheaded, but opened up his restaurant as per usual. Joe had a sudden cardiac arrest while scrambling eggs and could not be resuscitated at the scene. Obtained laboratory results demonstrated his potassium levels were below 2.0. The low potassium caused by un-monitored use of his diuretic likely caused his cardiac arrhythmia.

Furosemide can cure and can kill; only use when necessary.

Digoxin

Fairy Thimbles, Fairy's Glove, Witches' Gloves and Foxglove are previous names of this ancient remedy, the oldest of our medications. Digoxin was first named in the London Pharmacopoeia in 1650 and was brought into use by Dr. W. Withering in the treatment of dropsy (congestive heart failure) almost a century later. Although it is the oldest of our medications, Digoxin is controversial. Most recently, Digoxin was de-emphasized in the current guidelines for treatment of congestive heart failure where it had resided prominently for the last 400 years. (Chronic Heart Failure in the Adult: ACC/AHA 2005 Guideline Update for the Diagnosis and Management of (J Am Coll Cardiol 2005; 46:1116-43).

Digoxin is recommended for the treatment of fast rhythms associated with congestive heart failure. *Atrial fibrillation* is the rhythm causing your heart to beat irregularly and pump inefficiently. The top chamber of the heart is quivering in a chaotic manner and not contributing as a booster pump. The Digoxin helps slow this response. It is the only oral medication that increases the contractility- the strength of the heart. Digoxin increases calcium concentration within the cells by altering the function of sodium and potassium channels. Digoxin is a good medicine to add, even if the patient is not in atrial fibrillation. When a patient is still decompensated Digoxin can be added to help alleviate suffering.

The DIG trial was to answer questions relating to the use of Digoxin (The Digitalis Investigation Group. The effect of Digoxin on mortality and morbidity in patients with heart failure. N Engl J Med 1997;336: 525-33). The trial found the drug did not cause harm and could reduce the number of days hospitalized and alleviate patients' discomforts. Further examination of the data suggests the effect was somewhat less than previously reported. Data verifies women taking a normal dose had more adverse outcomes than women ingesting smaller doses. Women are smaller and the dose has to be down regulated for them. Digoxin has a narrow therapeutic index, which means small doses are beneficial, but too much Digoxin is harmful.

Sally Jean was an 84 year old female resident of the Older than Hills Nursing Home. She was a spinster survived by one sister and a number of nieces and nephew's that would visit her at rare times. She became increasingly frail and was now confined to a wheelchair. Her existence consisted of getting up in the morning getting partially dressed and being transported to the communal dining room for breakfast. She would engage in conversation with other residents reminiscing about days gone by and how different the world is today.

Upon retiring one evening, she was awoken by sudden fearful feeling, racing heart and terrible thumps within her chest. Although she had an irregular heartbeat for the last 20 years, the sense of strangling and smothering was a new event. The nurse was at Sally Jean's bedside within fifteen minutes and arrangements were made to transport her to the local emergency room.

Upon arrival to the emergency room Sally Jean was breathless and in a panic. Her oxygen saturation was 88% instead of 100% indicating her oxygen level was very low and her vital signs revealed an irregular heart rate that was faster than 150 to 170 beats per minute. Her respiratory rate was 30 per minute and blood pressure was elevated at 160/70. The emergency room physician recognized her rhythm as atrial fibrillation and controlled her heart rate with intravenous Metoprolol 5 mg given every 5 minutes for the next 15 minutes. The Metoprolol worked efficiently and in those 15 minutes Sally Jean's heart rate fell from 170 to just under 100 beats per minute. The fall in her heart rate was accompanied with an improvement in her panic and respirations.

The next morning the attending physician found Sally to be completely asymptomatic - still in her arrhythmia but at a slower rate. She had been loaded on a total of 1 mg of Digoxin during the night and her only medication at present time was Digoxin 0.125 mg a day. Her resting heart rate was 68 beats per minute. She was smiling and back to her baseline despite the irregular rhythm.

The attending physician now had to choose between rate or rhythm control and decide upon the benefits and hazards of anticoagulation therapy with 'rat poison' Coumadin. Sally was frail and already demonstrated bruises from previous misadventures when she had missed the toilet seat by a mere 3 inches. Since she was completely asymptomatic the attending physician decided to discharge her on a single medicine Digoxin 0.125 mg daily.

While other physicians may opt for another route, he felt the risk of hemorrhagic stroke and adverse events due to fall exceeded the patients benefit of anticoagulation therapy with Coumadin. The criticism is based on population trials demonstrating the benefit of Coumadin even in the elderly population. The study did not include very many people over 80 years of age. She didn't need any other rate controlling medicines because her activity level never required an elevated heart rate. Knowing the patient and being familiar with her heart proved to be a great advantage. Sally lived without any further episodes of shortness of breath and was able to visit her nieces and nephews for three additional years before an unfortunate fall caused a fractured hip and her demise.

Angiotensin-Converting Enzyme Inhibitors (ACEI)

One of the methods used to control blood pressure and volume status is the renin-angiotensin-aldosterone axis. These neurohormones allow us to survive hemorrhages, salt deprivation and dehydration. Captopril is of the first class of agents used to counter these neurohormones. In heart failure, this endocrine axis becomes very active because of reduced blood supply to the kidney. The neurohormones contribute to salt retention and fluid retention aggravating heart failure. The patient thus experiences edema and shortness of breath. The blood vessels become constricted making it more difficult for the heart to pump into small vessels as compared to large vessels. ACE inhibitors given in small doses gradually increase the size of the arterioles making it easier for the blood to eject into the vascular system. These medications need to be titrated from a small dose to a large dose to obtain the maximum benefit. Some of the ACE inhibitors that have been developed are: Captopril, Lisinopril, Vasotec, Accupril, Altace, Benazepril, Quinapril, Perindopril, Ramipril, Monopril, Mavik and Zestril. Although these inhibitors exhibit similar properties, they differ in dosage, duration of action, and how they penetrate into cellular membranes. These medications will prolong life in heart failure, improve cardiac function, and prevent the heart from remodeling into a sphere. Their actions include hemodynamic vasodilatation allowing the heart to pump with less work. These drugs work at the cellular level and help repair damaged cells and prevent sick cells from dying. Drugs with better cellular penetration may be better. The higher the dose the better, but titration with other drugs may need a reduction of the ACEI to add more beta-blocker.

One of the troubling side effects of these medications is the development of a cough. The cough is due to an increase in bradykinin. This is a beneficial substance that is cardioprotective, but directly produces a cough in some susceptible individuals. In treating heart failure we like to use a generic medicine such as Lisinopril and increase the dose 2.5 mg every week. By gradually increasing medication, patients are able to tolerate this blood pressure medicine even when their blood pressure is quite low. The goal of this medication is 20 to 40 mg a day. Patients who take their blood pressure will note that they have some low reading. The definition of too low a blood pressure is the inability to stand up and a decline in urinary volumes. The time to take blood pressure is when you are too weak to get out of the chair. The simple answer is if the patient can stand and are still able to urinate, then the blood pressure is adequate even if the blood pressure reads low.

Albert was a 57-year-old male with aortic insufficiency that had been present for nearly 30 years. The original cause of his aortic insufficiency was likely a bicuspid aortic valve. He had been a hard-working construction worker until he became disabled five years ago due to his ailing heart. His family had been lost somewhere in the past due to his requirement to move with the job.

Uncontrolled hypertension combined with the congenital abnormality caused his aortic valve to leak. Albert's heart deteriorated into extremely large ventricle with poor systolic function. His ejection fraction, the measure of the squeeze of the heart, was 12%. Normal is 60% and the time for surgery is when the ejection fraction falls to 50%. He had congestive heart failure for the last five years and it was managed only through diuretic therapy. His blood pressure was no longer hypertensive; in fact it was now hypotensive with blood pressures running 80 to 90 systolic. He was admitted to the coronary care unit for hypotension and pulmonary edema - a lethal combination known as **cardiogenic shock.** Cardiogenic shock has 60% mortality and is defined by low blood pressure and an inability to perfuse organs. To his benefit, his creatinine, a measure of renal function, was only mildly impaired. Nitroprusside was chosen as his initial therapy to improve his cardiac output. The CCU nurses initially refused to give the Nitroprusside because of his hypotensive status since their nursing book states that Nitroprusside lowers blood pressure and lowering the blood pressure would harm their patient. The cardiologist entered the room and alleviated their apprehension by remaining at the bedside and adjusting the intravenous Nitroprusside every three minutes doubling the dose from a very low initial starting value. Within 10 minutes the patient began to breathe much easier and to the amazement of the CCU nurses his blood pressure had climbed from 85 to 100 systolic. The patient's skin, which was initially icy and similar to a corpse, began warming to ambient temperature.

The next day, 3 mg of Captopril was started three times a day with the dose being adjusted upward with every other administration. Albert's blood pressure remained 100 systolic despite being titrated to 100 mg three times daily. His symptoms of congestion resolved. He was able to leave the hospital and was seen for three additional months in the clinic. Albert had made progressive improvement. His ventricle, however, only improved to an ejection fraction of 25%. Aortic valve replacement would not have relieved symptoms. Albert did very well for two years, disappeared, and was lost to further follow-up.

Angiotensin Receptor Blockers – (ARB)

Angiotensin II, an intense vasoconstrictor, is the end product of the rennin-Angiotensin system. One method to block the vasoconstrictor effects of Angiotensin II is with an analog of this substance. The analog sits on the receptor and has no effect on the cells and prevents Angiotensin II to attach to the cell. This blockade does not allow the vessel to constrict. Its' presence blocks the Angiotensin II from doing its' job and allows the arterioles to remain dilated. The class of (ARB) drugs include: Atacand, Avapro, Benicar, Cozaar, Diovan, Micardis, and Teveten. These drugs may be as effective as the Angiotensin converting enzyme inhibitors but they do not increase bradykinin. While there is no side effect of a cough, these drugs lack the potential benefits of stimulating Bradykinin. ARB's can reduce mortality when ACE-inhibitors can't be used and seem to provide an additional benefit when added to ACE-inhibitors. ARB medications appear to work well with elderly patients since they self titrate. Patients tend to complain less because blood pressure does not have as great shifts during titration.

Edna was an 83 year-old retired school nurse who was not quite 5 feet tall. During her life she only weighed more than 100 pounds when pregnant. She currently weighed 93 pounds. She still lives by herself and enjoyed gardening "chopping." She developed heart failure after an inferior myocardial infarction three years ago. The posterior leaflet of her mitral valve was now tethered forward resulting in leak into the left atrium. This leak would increase the pressure in her lungs and caused her to be short of breath. Surgery was a consideration but with her small stature the risk of procedure was considered relatively high and Edna did not wish to pursue this high-risk venture. For this reason was placed on medical therapy. She did not tolerate the initial titration of ACE inhibitors because her blood pressure would fluctuate approximately 1 hour after taking her medications. Her blood pressure would plummet even further when she would stand. The transient hypotension resulted in several falls but to the amazement of all her caretakers she never suffered a fracture.

Edna's ACE inhibitor was changed to an Angiotensin receptor blocker given at night. She was initially given a very low dose that was slowly increased each week. Her breathing improved and the shortness of breath became a distant memory inspiring her to plant a garden and was looking forward to picking up pecans in the fall.

Spironolactone

Spironolactone is one of the oldest antihypertensive agents. This drug blocks Aldosterone. Aldosterone was discovered over 50 years ago in the adrenal gland. Most cells of the body particularly the heart, brain, and kidney produce Aldosterone. It causes the kidney to retain sodium and is beneficial when deprived of salt and water preventing dehydration. This substance is very deleterious in the condition of heart failure because Aldosterone adds additional edema fluid, causes fibrosis, stiffening of the heart, and increases inflammation. Other conditions in which the life saving hormone becomes an enemy include nephrotic syndrome and cirrhosis.

The potential benefits of this drug were studied in the RALES trial. It selected patients with significant heart failure class III and IV (good chance of dying in the next three years). There was a 30% reduction in deaths with the use of Spironolactone. Spironolactone is cheap and costs $9 for a month's supply. It has the side effect of gynecomastia (breast growing and breast tenderness) in about 10 percent of patients. Potassium can become elevated when taking this drug and high potassium foods need to be avoided and all potassium supplements stopped. Eplerenone is a Spironolactone-like drug that acts in a similar manner by inhibiting Aldosterone but does not have the estrogen effects. Breasts do not get tender and enlarged. The price for this pill is $4 a day. Two percent of the 10 percent with breast tenderness will agree to pay the extra money to avoid the discomfort. Eplerenone showed a benefit in the Ephesus trial in post myocardial infarction patients with heart failure and ejection fraction of 40 percent. The drug will allow patients to live a longer life.

Although Spironolactone is over 50 years old, it has properties that are just beginning to be discovered. Research shows it may decrease fibrosis within the myocardial cells and within the vasculature. The drug likely has some anti-arrhythmic properties in that it prevents sudden death by keeping potassium levels elevated. The drug reduces inflammation and the combination of its' beneficial properties makes this an ideal medication for the treatment of congestive heart failure.

It also benefits patients with *recalcitrant hypertension*. These are individuals whose blood pressure remains elevated despite the use of three different types of hypertensive medications. By adding a low dose of Spironolactone to these patients, blood pressures can be quickly brought under control. Spironolactone will likely become the drug of choice for stiff hearts that have become scarred with fibrosis from old heart attacks or hypertension. Studies to show the benefit in individuals with diastolic dysfunction and stiff hearts are still pending.

There is a price to pay for so many wonderful beneficial properties. The price is the deleterious affects of hyperkalemia which is the marked elevation of potassium in the blood. Intravenous potassium can be lethal and is the primary agent used in execution of prisoners who have committed horrible crimes. The inmates are given sedative medicines followed by a lethal dose of potassium. Elevated potassium's slightly above normal is a good thing in patients. Low potassium levels are positively correlated with high blood pressure and arrhythmic deaths.

I will use the term "watermelon syncope" to explain this phenomenon. The job of Spironolactone is to excrete sodium, which is deleterious in heart failure, and hold on to potassium. This gives great benefit to patients who are hypertensive and patients who require diuretics that lose potassium. In treating heart failure patients we use multiple medications that retain potassium. The ACE inhibitors, ACE receptor blockers, and even Beta-blockers tend to raise potassium levels. Once the patient has been stabilized in their heart failure, potassium wasting diuretics are discontinued. The normal range of potassium ranges from 3.5 to 5.5. With the use of these medications potassium levels frequently can run elevated above 5.5 to 6. When potassium levels become greater than 6.5 deleterious affects begin to occur. Watermelon season then arrives, or any other season that has foods containing increased potassium. The patients know they're not allowed to use salt so they use a substance called 'no salt.' Unfortunately, "no salt" substances are derived from potassium. Although watermelon does not have a large quantity of potassium, when you add potassium salt to the watermelon, the intake of potassium can become excessive. As the potassium levels rise above the allowable level of 6.0, the heart rate decreases, resulting in less blood flow to the kidneys. The kidney retains more potassium because of the reduced blood flow. This cascade of events results in elevated potassium levels. A high level of potassium produces a lower heart rate potentially causing extreme weakness or spells of passing out in the patient. The proper use of this medicine requires patients to be aware of foods high in potassium. Previously, treating heart failure patients included recommending patients eat several bananas a day to replenish potassium. In the old days, simple diuretics were the mainstay of treating congestive heart failure and hypokalemia (low potassium level) was very common. Heart failure medication used today can result in watermelon syncope when combined with foods high in potassium. Below is a list of foods high in potassium that are harmful when taken in excess.

Fruits – Apricot, Avocado, Banana, Cantaloupe, Casba, Dates, Dried Fruit, figs, Honeydew, Mango, Nectarine, Orange, Juices of the fruits, Papaya, Plums, Prunes, Raisins, Rhubarb
Vegetables – Artichokes, dried Beans, Broccoli, Brussels sprouts, Celery, Endive, Greens, Kale, Kohlrabi, Lentils, Legumes, Lima Beans, Mushrooms, Parsnip, Potatoes, Tomatoes,
Other Foods – Bran, Coffee (limit 2 cups/day), Chocolate, Coconut, Granola, Ice Cream (limit 1 cup/day), Molasses, Nuts/seeds, Orange flavored pop, Salt Substitutes, Snuff/chewing tobacco, Tea (limit 2 cups/day)

If the potassium level is elevated, a history of excessive intake of foods from the above groups should be obtained from the patient or family members. If a patient has hypertension and is not on Spironolactone or Eplerenone, eating from this list of foods will cause potassium levels to rise thus lowering blood pressure.

George Sinclair was a 73 year-old retired banker who suffered a myocardial infarction with resultant heart failure. He was on a beta-blocker, ACE inhibitor and large doses of Lasix to maintain his fluid balance. His serum level of potassium ran 3.2 to 3.5 on 40 meq of potassium two times a day. George is still suffering from peripheral edema and wears bedroom slippers rather than shoes due to swelling. He continues to use his salt shaker and drinks what ever he "Darn well pleases." He has a component of right heart failure which is the best recipe for continued edema and bedroom slippers.

After visiting the heart failure NP (Nurse Practitioner) Tracy in the heart failure clinic, his potassium was discontinued and Spironolactone was started initially at 25 mg a day and then raised to 50 mg. With this medication and better compliance, his peripheral edema resolved. Three weeks later George became lightheaded and dizzy with a low home health blood pressure of 80 mg millimeters mercury. His primary care physician, lowered his dose of ACE inhibitor.

The following week he came to the heart failure clinic to see NP Tracey and was feeling fatigued and still short of breath. She discontinued the diuretic therapy with Lasix and instructed George to increase titration of the ACE inhibitor again in one week. George was told that he would likely gain 7 pounds of weight. Patients who continue to take Lasix with no excess fluid in their body develop dehydration. The symptoms of dehydration included lightheaded dizziness, fatigue and paradoxically shortness of breath. The shortness of breath comes from low cardiac output and ventilation perfusion mismatches. These are fancy words for not enough blood to fill the lungs. Everyone experiences this if they sweat too much. George's expected weight gain was 7 pounds based on his previous dry weight. He had finally listened to Tracey and became salt and water compliant and no longer needed the Lasix. The Spironolactone, which was a weak diuretic, helped mobilize the peripheral fluid back into the vascular space. This medication will not significantly dehydrate a patient on its own. Two weeks later, George Sinclair was back to exercising and decided to volunteer his services by teaching a class on personal investments.

Sylvia is a mildly obese 58 year-old waitress. Hypertension tended to run in her family and a sister, her mother and one brother died from the complications of blood pressure elevation. Her brother was on renal dialysis for six years until he died. Both her sister and mother suffered cerebral hemorrhages secondary to uncontrolled hypertension. She understood taking medication is essential to prevent similar life tragedy. Sylvia was frustrated that despite all of her efforts blood pressure was still elevated. She attempted to control her salt intake by never adding any salt to any of her meals and was therefore frustrated that despite her efforts she still had an elevated blood pressure.

Her current medical regime consisted of Lasix, an ACE inhibitor, a beta-blocker, and a calcium channel blocker. A good blood pressure for her was 150 systolic and many times under a little stress she would record values of 200 systolic. Spironolactone was added at 25 mg a day. Potassium supplementations were discontinued and dietary instructions were completed. The next week her blood pressure was 135 systolic.

She had previous evaluation of her renal arteries with a magnetic resonance imaging. She had undergone blood tests for pheochromocytoma, renin and Aldosterone levels. These studies were within normal limits. Despite the blood tests demonstrating no evidence for Aldosterone excess, Sylvia responded nicely with normalization of her blood pressure. Conn's syndrome, a secondary cause of hypertension, is a rare finding in less than 1% of hypertensive patients. It is an Aldosterone producing adenoma and removal of this small tumor will cure hypertension. Hypertension due to Aldosterone excess not related to an adenoma is much more common. The best screen for this test is to give Spironolactone and demonstrate blood pressure improvement. In treating difficult to control hypertensive patients avoid expensive tests and try a medication that costs $9 a month.

Sylvia completed her undergraduate degree and became a social worker helping patients obtain medications to treat their hypertension, diabetes, and heart failure.

Beta Blockers

Beta-blockers play a prominent and interesting role in the treatment of congestive heart failure. For many years it was a contraindication (meaning it was harmful) to give patients suffering from congestive heart failure beta-blockers. These medications were known to decrease contractility (the squeezing power of the heart) and were felt to be harmful in this condition. As Dr. Houck has already warned our readers, not everything that he was taught is correct. A group in Sweden persisted in giving beta-blockers to congestive heart failure patients and it took nearly 20 years to prove that their position was correct. (K. Swedberg, *History of Beta blockers in congestive heart failure*, Heart. 1998 June; 79(Suppl 2): S29–S30) Beta-blockers are now considered an absolute indication for all forms of congestive heart failure. The medical community has completed a 180° turn. Numerous clinical trials show improvement in ejection fraction and mortality when beta-blockers are given to individuals with congestive heart failure. The exact mechanism of how beta blockade works is still a mystery. There are currently three drugs that have demonstrated clinical trial benefit in the final stages of congestive heart failure Bisoprolol (Zebeta), Carvedilol (Coreg), Metoprolol succinate (ToprolXL).

A number of neurohormones are activated to help preserve blood pressure when the failing heart does not produce adequate blood supply to the body. The sympathetic nervous system is highly active in congestive heart failure. These catecholamine hormones produced by the adrenal gland are the same hormones that become activated during the "fight or flight" reflex. These hormones increase heart rate and cause the blood vessels to become constricted in an effort to preserve blood pressure. These same hormones also increase the contractility of the heart making it squeeze harder than it should and this sympathetic activity tends to cause a burnout of the myocardial cell. The cell in an effort to preserve itself will ignore these hormones and become weak. By blocking the effect of these hormones with the beta-blockers the heart function can then regain function.

The heart rate longevity model can be applied to beta-blocker therapy. The small little shrew, which has a heart rate approaching 300 beats per minute, has a very short life span of less than two years. The elephant, the human, and whales have heart rates approaching 70 beats per minute allowing us to live approximately 70 years. The Galapagos turtle, which lives over 300 years, has a heart rate of six beats per minute. The slower the heart rate the greater the longevity. Beta-blockers decrease heart rate and in simple terms increase longevity.

Dr. Houck maintains the theory that beta-blockers do not work on the receptors and enzymatic pathways but promotes the regeneration of cardiac cells by either increasing the number of circulating stem cells or increasing stem cell proliferation within the myocardium. The antioxidant effect of Carvedilol may be the reason it stands out among other beta-blockers. This medication tends to improve cardiac muscle at lower doses and often is the preferred medication for treatment of congestive heart failure.

The time course for improvement in beta-blocker therapy is a lengthy one. Individuals have to be titrated up in their medication, which may take 6 to 8 weeks. Approximately 3 to 4 months after the medications are titrated up; there is marked improvement in the individual's heart function as measured by the ejection fraction. Three to four months is the length of time that is required for new cells to take the place of old and dying cells in the myocardium to rejuvenate the heart. We have observed individual's hearts have their function improve from a suffocating 20% ejection fraction to a normal 60%. If for some reason their medications are stopped, their cardiac dysfunction returns. It is unlikely that the activation of the neural hormones were acting at the cellular level. It is very likely that the beta-blocker therapies help rejuvenation of the myocardium and stopping the medications allows the heart to age into disrepair.

Carvedilol is a beta-blocker that also has alpha blockade and antioxidant properties. The medication needs to be started at a very low level and titrated up each week. The goal therapy is at least 6.125 mg twice a day to obtain benefit from the medication. The larger the dose is titrated the more benefit is obtained. For patients weighing more than 180 pounds, the recommendation is to titrate to 50 mg twice daily. For fewer than 180 pounds 25 mg twice daily. The medication needs to be taken with food because if it is taken on an empty stomach the pill will be rapidly absorbed and blood pressure falls just as rapidly with resultant symptoms of fatigue, dizziness, and ill feeling. This medication by far gives the greatest improvement in ejection fraction when added to the other heart failure medications.

Carvedilol is one of the most prescribed medications. The medication is considered a miracle drug but Dr. Houck does not agree with the company marketing the medication as an immediate post-myocardial infarction medication. (The Capricorn Investigators. Effect of Carvediol on outcome after myocardial infarction in patients with left ventricular dysfunction: The Capricorn Randomized Trial. **Lancet** 2001;357:1385-1390) used Carvedilol post-myocardial infarction in a setting of modern medical therapy including ACE inhibitors. The trial was a placebo-controlled trial. It is unfortunate the trial did not include a comparison to another beta-blocker that has been shown to be of benefit post myocardial infarction. Beta-blockers were one of the first medications that demonstrated a post-myocardial infarction mortality benefit. The argument counters the beta-blockers were not studied in the setting of current medical therapy that included ACE inhibitors and may not have been used in patients with poor left ventricular function. The survival curve demonstrates benefit of Carvedilol over placebo beginning at the fourth to fifth month post-treatment. In the first four to five months post-myocardial infarction there may have been more deaths in the Carvedilol group than in the placebo group. These deaths accounted for nearly 25% of the deaths during the study. There is not any beta-blocker trial prior to this trial that demonstrated early harm. In a recent trial of high dose Metoprolol post acute myocardial infarction demonstrated more cardiogenic shock. The study of early harm and late benefit is not a statistical test. A brief discussion of survival curves is needed to make this point

Survival Curves

What can survival curves teach us? What have we been taught to ignore? When trials are performed pre-specified endpoints are set as a safety margin. If a therapy crosses this boundary, there is statistical evidence that harm or benefit has occurred and the trial should be stopped. We have been taught to ignore the beginning of the curve, until an appropriate number of events have happened since chance may be the reason for those early changes. This reasoning is valid when one considers a therapy to have only a single result and that the population is uniform in its response. If the therapy is bimodal (early harm with late benefit or early benefit with late harm), ignoring the beginning of the curve, or not extending the curve may lead to the wrong conclusions. If the population is not uniform but has characteristics that respond differently to the therapy, ignoring the early and late curves may lead to an inaccurate confirmation or rejection of a hypothesis. Post trial analysis with appropriate subgroups could potentially identify these groups who receive harm or benefit. There would have to be a biologic difference that would explain the harm or benefit in the post-test analysis. Examples are given to illustrate the problem.

The V-HeFT I in the 1980s was a pivotal trial in vasodilator therapy for congestive heart failure (Cohn JN, et al. Veterans Administration Cooperative Study on Vasodilator Therapy of Heart Failure: influence of prerandomization variables on the reduction of mortality by treatment with hydralazine and isosorbide dinitrate. Circulation. 1987 May;75(5 Pt 2):IV49-54.) It took 20 years until a post trial analysis demonstrated that most of the benefit of the trial came from the hypertensive African American population. A-HeFT proved in African Americans that hydralazine nitrate combination was very useful. The original V-HeFT I population was bimodal in that it had a population of coronary disease and a population of hypertensive disease. The combination of hydralazine and nitrate is very effective in hypertensive disease. The biological explanation is the nitric oxide pathway that is abnormal in the black population is treated very effectively with hydralazine and nitrate. The posttest analysis of V-HeFT I was accurate and confirmed with the A-HeFT trial.

Thrombolytic therapy for acute myocardial infarction is used as an example of a non-uniform population with early harm and late benefit. Within the first 24 hours of therapy there appeared to be harm with the treatment arm curve dipping below the placebo curve. The curve rapidly improved and demonstrated significant benefit. The trial was not stopped because the events were low in numbers and did not cross the boundary for stopping the trial. At the end of the therapy there were more people alive with treatment. The population, however, had a subgroup of patients who had tendency for intra-cranial hemorrhage. These individuals died due to their predisposition to hemorrhage and suffered early harm from the thrombolytic agent. If a large number of individuals were at risk for intracranial hemorrhage, the trial would have been stopped and we would never have seen the benefit from the population who was not at risk. A Study of the population who had hemorrhage could identify individuals at risk and give biologic plausibility of the early harm.

The COMMIT trial that used Metoprolol 15 mg IV followed by Metoprolol 50 mg PO q 6 hr then Metoprolol succinate 200mg daily can be used as an example of early harm with late benefit. (Chen ZM, et al., Early intravenous then oral metoprolol in 45,852 patients with acute myocardial infarction: randomised placebo-controlled trial. Lancet. 2005 Nov 5;366(9497):1622-32) There was a significant reduction of re-infarction and ventricular fibrillation but a greater incidence of cardiogenic shock during the first day. The dose of beta-blocker for the trial may be the reason for early harm. Previous data with lower doses of beta blocker did not show this early harm. There may be an additional reason for this result. The population was not uniform and had more severe Killip classification than previous myocardial infarction trials. The Killip classification rates myocardial infarction I - IV based on their clinical presentation and predicts mortality.

The current drug eluding stent debate is an example of early benefit with late harm at least in a portion of the population that suffers late stent thrombosis. The survival curves were not extended to find this possibility.

The shape of the survival curve of the treatment versus placebo can be used to postulate mechanisms of action. I personally observed this while sitting on the safety committee for the AFCAPS-TXCAPS primary prevention trial when it was observed that the curves separated early in the course of therapy within 3 months. The better result was not due to the plaque shrinking but something that still remains a mystery. The time course of separation of the curves can potentially be used to determine a biochemical, anti-inflammatory or cellular regeneration etiology. The benefit would likely be seconds to days with a biochemical or enzymatic effect, days to weeks for an inflammatory response, and months for cellular regeneration.

The CAPRICORN trial will illustrate the potential importance of the shape of the curve in the following slide provided from industry. This curve appears to demonstrate an early steep drop suggesting early harm with a very good late benefit. The crossover to benefit is 4 to 5 months after the myocardial infarction. An explanation for the early harm could be suggested from the COMMITT trial with patients having hypotension from the alpha-blocker component or inadequate beta blockade to prevent re-infarction or ventricular fibrillation. The fascinating late benefit of this drug at 4 months is similar to what is observed in treatment of heart failure and could suggest cellular regeneration as the mechanism.

Before jumping to conclusions, this curve is not colored and it is impossible to predict which curve represents placebo in the less than 3 months. Data presented in the paper does not allow one to know for sure which curve is the placebo. In the Lancet article the curve is colored so that Carvedilol always behaved better than placebo. If this is true, there is no explanation of the discontinuity where the curves cross. It would suggest a bimodal population or bimodal response in both the placebo and treatment limbs. This may be a statistical phenomenon or a misprint. I have never been able to learn the truth.

Survival curves are not simple. They contain much more information then utilized. A bimodal shape to the curve could suggest early harm with late benefit or a non-uniform population. The time lag of benefit could point toward a biologic mechanism. A strategy to analyze survival curves after a trial should be incorporated into the original protocol. Trials that have already been performed should have the survival curves examined especially if a biologic explanation for a bimodal response is reasonable.

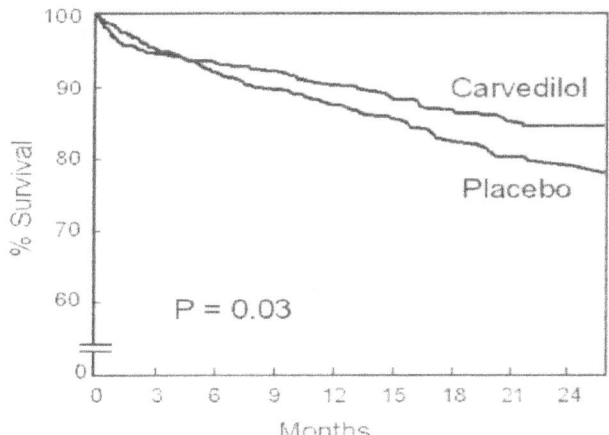

This curve was given to me by Drug Representative to help my confusion

Carvedilol causes more hypotension than a beta blocker without alpha blockade. It has to be given in smaller doses. There is high likelihood that the hypotension caused by the alpha blockade caused early harm as well as the small dose allowing arrhythmias. Both conditions are biologically plausible explanations for early harm. For these reasons, consider not using Carvedilol for the first three to four months after myocardial infarction, but initiate standard beta-blocker therapy with Metoprolol, and then change to Carvedilol beta-blocker at three to four months if heart failure persists.

Dennis, a 38 year-old male, is a tall slender, well-tanned good-looking young man who has diabetes. He enjoyed doing the manual labor demanded by his irrigation business, but found it became exceedingly more difficult. He became a supervisor in his own business and watched other laborers complete tasks at his wishes. Shortness of breath had continued but he was not concerned until he noticed increased swelling in his legs and difficulty sleeping at night. He was evaluated in the cardiology clinic and underwent cardiac catheterization that demonstrated some mild coronary disease but nothing that would explain his heart dysfunction. An echocardiogram demonstrated his ejection fraction was 18%. At the time he was first diagnosed the standard of care was Digoxin, ACE inhibitors, and diuretic therapy. Nitrates were given at night to relieve breathing difficulties. He continued with progressive symptoms and arrived in the cardiology clinic with mild hypoxemia and air hunger. He was admitted to the family practice service for intravenous therapy of this congestive failure. Coreg had recently been approved for the treatment of heart failure. At that time, giving the medication in the hospital was prohibited. Despite this warning, Coreg was initiated at 3.125 mg twice a day. Dennis tolerated the medication and was followed in the heart failure clinic for chronic titration to 50 mg twice daily over the next three months.

At six months of therapy Dennis fired his laborers and went back to manual labor digging ditches for his irrigation systems. A repeat echocardiogram demonstrated complete resolution of his heart failure with ejection fraction of 50%. After two years Dennis decided to discontinue his medications because he was feeling well and was burdened by the financial pressures of Coreg. The symptoms again returned and his ejection fraction fell from 50% to 20%. He returned to the heart failure clinic where his titration of medications was again performed. To his great luck, his ejection fraction again returned to normal. In the future Dennis will not stop his medications for any reason.

Titration of these medications is dependent on patient tolerance, patient psychology, blood pressure, and renal responses. The scheme presented should work in nearly everyone because it goes slow changing medications weekly. The slow titration allows the body to adjust to the changes. The blood vessels dilate slowly allowing preload to shift to the afterload. The heart becomes happier and happier as the blood vessels get larger and the heart shrinks in size. With a happier heart, the muscle then begins to remodel, regenerate and further strengthens the heart. In some patients the heart will be completely new and healthy.

WEEK	FUROSEMIDE	DIGOXIN	ACEI	ALDACTONE	COREG
	AM / PM	PM	AM / PM	AM	AM / PM
1	40 /	.25	/ 2.5	25	
	LAB BUN/CREAT/ E-GROUP 4 DAYS AFTER STARTING				
2	40 /	.25	2.5 / 2.5	25	
3	20 /	.25	2.5 / 5.0	25	
	LAB BUN/CREAT /E-GROUP				
4	20 /	.25	5.0 / 5.0	25	
5	/	.25	5.0 / 5.0	25	3.125 / 3.125
	LAB BUN/CREAT /E-GROUP				
6	/	.25	5.0 / 7.5	25	3.125 / 6.25
7	/	.25	7.5 / 7.5	25	6.25 / 6.25
	LAB BUN/CREAT /E-GROUP				
8	/	.25	5.0 / 7.5	25	6.25 / 9.125
9	/	.25	7.5 / 7.5	25	9.125 /9.125
	LAB BUN/CREAT /E-GROUP				
10	/	.25	7.5 / 10	25	9.125 / 12.5
11	/	.25	10 / 10	25	12.5 / 12.5
	LAB BUN/CREAT /E-GROUP				
12	/	.25	10 / 12.5	25	12.5 / 15.625
13	/	.25	12.5/ 12.5	25	15.625/15.625
	LAB BUN/CREAT /E-GROUP				
14	/	.25	12.5/ 15	25	15.625/ 18.75
15	/	.25	15 / 15	25	18.75 / 18.75
	LAB BUN/CREAT /E-GROUP				
16	/	.25	15 / 17.5	25	18.75/ 21.875
17	/	.25	17.5 / 17.5	25	21.875/ 21.875
	LAB BUN/CREAT /E-GROUP				
18	/	.25	17.5/ 20	25	21.875/ 25
19	/	.25	20 / 20	25	25/ 25
	LAB BUN/CREAT /E-GROUP				
20	/	.25	17.5/ 20	25	25/ 25
21	/	.25	20 / 20	25	25/ 25
	LAB BUN/CREAT /E-GROUP				

The entire titration scheme takes 21 weeks with blood testing every two weeks to check kidney function and potassium. Doses need to be altered for renal failure and the titration scheme may have to be slowed down due to patient tolerance or speeded for hypertension. It should be noted that diuretics are discontinued early to allow for further titration of the medication. A common problem is patient dizziness due to low blood pressure from too much diuretic. The mistake doctors make is to stop the life prolonging ACE inhibitor or the beta-blocker instead of the life shortening Lasix. In patients over 180 pounds the goal of Carvedilol titration is 50 mg bid.

After explaining the heart failure medications to Murray, Tracy replied "Murray these are your instructions and I will be seeing you frequently in the beginning to listen to your complaints and act as a cheerleader. You will be following medication instructions, fluid restriction, and you will not let your tongue taste salt. Do you have any questions?"

Murray, hesitated, coughed, and looked into Tracy's brown green eyes. After an uncomfortable period of silence, Murray spoke. "I am not sure how to express myself; I am frightened of not being able to breathe, and more frightened of becoming dependent on others to get through the day. I am old and recently feeble, but my heart looks to the future and new adventures. When not clouded by hypoxia, my mind is as clear as it was 60 years ago. I trust you and your knowledge and will do my part in taking medications. I don't want to go quietly and want to raise a ruckus. I still have desire, but lack ability. I want to have one more adventure and I want you to be part of this old man's life. I am giving you 6 months to repair my poor heart and then I want to make a trip to the Philippines to rediscover a mystery I found during World War II. I want to invite you to be part of my expedition."

Tracey's mouth fell open and was momentarily lost for words. At that moment Murray seemed at least 40 years younger. He had an appeal that was difficult to describe. She had similar comments from other male patients, but this one seemed different. It was genuine and completely out of left field. She simply replied, "Sure, see you next week."
As Murray left the room, he turned to Tracey and said he was going to hold her to her promise and was going to recruit Dr. Collette, Dr. Houck, and anyone else he fancied. The adventure would be hot, full of smells, exercise, and would reveal a great mystery if his memory from long ago was not tainted by years and a bout of Dengue Fever. He had eluded the Japanese soldiers, but a small mosquito hungry in the early evening had infected him with bone break fever. He remembered the tiny red dots across his arms chest and legs. He could not move due to the pain in his bones and overwhelming fatigue. Murray did have hallucinations from this illness, but his adventure to the ancient hidden cave was too vivid to be a hallucination. This cave after all had saved his life. He always wanted to go to see if he could find the cave but always had other obligations that never allowed him to complete his adventure.

The next week Murray was at Tracey's door 10 pounds lighter with a grin and a new tan. He was eager for more health education and to boast that he had quit smoking and was now careful in his diet. Tracy was a bit overwhelmed by his improvement and enthusiasm to get better. A new life torch had been lit in the previously broken man. Spirit, desire, mental attitude keeps some patients going and allows others to fall into the gutter. She decided to teach him more about his medications and increase the rate of his medication titration. She wanted to ask more about his planned safari but was too timid to do so. Tracey just wanted the old man to get better. She was finishing her PhD and had not had a vacation for 4 years. She was too busy to date and felt she was getting very dull. She was ready to have a break. While daydreaming of far away tropical islands she restarted her education.

Amiodarone

There is only one anti-arrhythmic medication; at least, that is what Dr. Houck has expressed. This medication was originally developed to treat Angina, the pain in the chest that would occur with blocked coronary arteries. The drug was developed in Belgium in the 1960s and its anti-arrhythmic properties were not observed until after its release. It eventually came to America and was used for this purpose. The medicine was found to be very effective in treating atrial fibrillation, ventricular tachycardia and had little pro-arrhythmic affects (cause arrhythmias). It is a once a day medication with a half-life of months. A half-life refers how long it takes for a drug to decrease in the body. A long half-life means it takes many doses of the medicine to get to a certain level and a very long time to be eliminated from the body. It is a medication you could forget to take on vacation and there would still be enough working to protect you three weeks later. Most anti-arrhythmic medications have side effects and may cause sudden death. Amiodarone is well tolerated but does have significant side effects in rare individuals. While liver and lung fibrosis are the most serious consequences, thyroid disease may be caused by the large amount of iodine in the medication. Both hypothyroidism and hyperthyroidism can be caused by the same medication.

Despite all of the drawbacks, the medication is the only safe anti-arrhythmic medication. If the heart rate is too fast, Amiodarone should be used. If it is too slow, use a pacemaker. If it is both fast and slow, use Amiodarone and a pacemaker. If the patient has asthma and can't take a beta-blocker, Amiodarone can be used. If the patient has angina and is not better, Amiodarone should be used. It is a "go to" medication when all else has failed. Even though it has a prolonged loading time it is effective when given as an intravenous medication. Cardiac patients with significant problems will live longer on this medication. As soon as the medication is not required, it should be reduced in dosage or eliminated to avoid long-term side effects. Patients who do not have good long-term prognosis should not worry about side effects of this medication.

Ruth was a 73 year-old retired postal worker. The last five years she suffered with fullness in her head,—and progressive shortness of breath. She had a known cardiomyopathy with an ejection fraction of 35%. She also had complex ventricular ectopy implying she has worrisome additional heartbeats coming from the ventricular surface. She underwent electrophysiologic testing and was found not inducible for sustained arrhythmias. She had been placed on all the appropriate heart medications and demonstrated no improvement in her symptoms or her ejection fraction. During her echocardiogram the cardiologist noted that she had more extra redundant heartbeats than normal heart beats. Many of these heartbeats occurred at a relatively fast rate. Some of the PVCs were associated with canon "A" waves. Canon "A" waves occurred when the atrium contracted against the closed AV valve. This would cause increased central pressures. This was the symptom of fullness in her head that she had been experiencing for five years. Although her sinus node showed a heart rate of 70 beats per minute, her average heart rate was much higher. Her ventricle was working too fast with resultant inefficient contraction due to all the extra heartbeats. With each heartbeat there was a leak of the mitral valve that would further increase the pressure in her lungs.

the complex ectopy was noted in the cardiology clinic. The patient could not feel these palpitations. She would have occasional fullness in her head. Her main complaints were fatigue and shortness of breath due to her cardiac dysfunction.

Ruth was started on Amiodarone at 200 mg daily and was told to return in three months for re-check of thyroid and liver function tests. It took three months for the Amiodarone to reach its effective level. Ruth returned to the cardiology clinic free of symptoms. A repeat echocardiogram demonstrated complete resolution of her left ventricular dysfunction. Ruth was suffering from cardiac dysfunction secondary to an electrical abnormality. There were rebel cells in her heart. These cells wanted to take control of the heart. The resultant inefficient contraction resulted in cardiac dysfunction. By correcting these extra heartbeats, putting the rebels under house arrest, the heart was able to function normally. Ruth had to pay a price of continuous monitoring of her thyroid, liver function tests, and potentially pulmonary fibrosis. In two years she would have to take the Levothyroxine because of drug induced hypothyroidism. Ruth continued to do well and had her Amiodarone dosage dropped to one half-tablet daily with no further increase in her ectopy.

Statin's

"My great aunt's niece was on that Statin medication and her joints and muscles hurt. I don't want to take that medication. The TV says it is bad for your liver and it will cause muscle pain."

A vision comes to mind hovering over this type of patient. The vision is a Jackass spreading its hunches getting ready to defecate onto the patient. The Jackass is drifting on a cloud just above the patient. I frequently look up to see this vision when I encounter these difficult predestined to fail patients. Statin myalgias do occur in 7% of the population and is likely due to a rare metabolic fatty acid pathway that is poorly understood. CoQ-10 may help some of those patients. The television tells patients to check with their doctor if they have muscle aches. The real harm the television refers to is a rare condition, one per million patient years of statin use, known as rhabdomyolysis. Rhabdomyolysis is breakdown of muscle manifested by can't get out of the chair weakness and dark urine. This unfortunate condition was first identified in transplant patients taking anti-rejection medications where there was an interaction between medications. Are these medications safe? Compared to over the counter pain relievers like acetaminophen or NSAIDs, it is quite safe. Acetaminophen has caused many deaths due to over dosage in suicide attempts or misuse destroying the liver. NSAID's have killed more people than AIDS. Deaths due to Statin medications is quite rare.

The action of Statin's is more than lowering cholesterol. The benefit from taking these medications increases over time with a small benefit within the first year and a 51% reduction by the third year of use. Overall there is a 61 % reduction in heart attacks and a 17% reduction in stroke. The real benefit is likely under reported since the effects tend to give increasing benefit with the length of exposure. The studies only go for two to three years. Statin's have pleotrophic effects which means they do other things besides lowering cholesterol. These amazing medications will help people withstand surgery, reduce hip fractures, decrease vascular inflammation, increase circulating stem cells, and diminish the risk of dementia. In addition, survival with heart failure and transplantation is extended. Who would not want to take these medications – someone who would rather get pooped upon by a Jackass?

The 1985 Nobel Prize in Physiology or Medicine was awarded jointly to Michael S. Brown and Joseph L. Goldstein (nobelprize.org) for elucidating the regulation of cholesterol metabolism by LDL receptors. This study (Goldstein, J.L. and Brown, M.S., "Familial hypercholesterolemia: Identification of a defect in the regulation of 3-hydroxy-3-methylglutaryl coenzyme A reductase activity associated with overproduction of cholesterol" PNAS USA, 70:2804-2808, 1973) explained cholesterol metabolism and described a small genetically prone group of individuals who did not have LDL receptors. These homozygous effected individuals could have a heart attack by the age of 10 and would not live beyond their twenties. The elucidation of this mechanism gave birth to Pravachol and Mevacor the first HMG CoA reductase inhibitors, the first Statin medications.

There have subsequently been many Statins introduced into this billion-dollar industry. The Statins differ in potency, lipid solubility, P450 metabolism, and price. Doctors believe that they have a class effect, which means they all work about the same. However, this theory remains untested in research studies.

Mabel was a 77 year-old female who lived just outside a small farming community. She always wore makeup and did not feel proper without high-heeled shoes. Her physician had recommended that she quit wearing those shoes because he was concerned that her five-foot frame of 240 pounds would result in a malfunction of the shoes and a terrible fall. Despite her size, she moved with grace and was considered a fine and respected lady of the community. She had hypertension and terrible lipid profile with and LDL cholesterol of greater than 160.

Her cardiologist and primary care physician attempted to get her on a Statin medication. All of the Statins were tried but within 6 weeks a new pain or ache would occur bringing a halt to the medication. Muscle enzymes were never elevated. She even failed Zetia, a drug that did not have any muscle influence. Red yeast rice may have been acceptable. Metamucil caused gas, and Benacol spread caused bloating and diarrhea. Everything caused this patient some symptom.

Her last visit also demonstrated an elevated high sensitivity CRP (C reactive protein) of 7, which meant that in addition to her abnormal lipids, she had inflammation giving her high risk of myocardial infarction and stroke.

Two months later she awoke in the early morning after a thunderstorm with a crushing indigestion type of pain combined with nausea and sweating. She went to the local ER 12 hours into her pain and had her pain relieved with one Ntg. Ntg is Nitroglycerine which is a component of dynamite that causes coronary vasodilatation and is frequently given in a sublingual, under the tongue form. EKG and enzymes were positive and she was transferred to a higher level of care for a cardiac catheterization. Mabel was fortunate that the occlusion was not in a main branch but in a very distal small vessel resulting in a very small myocardial infarction. The same Satins that she could not take were now tolerated. She knew she would have aches and pains but she did not want to experience another heart attack. Her cardiologist was wise enough not to say "I told you so," but he sure did have those words on the tip of his tongue. He did congratulate her for sticking with the drug and encouraged her to continue.

To save money we split pills. For example, a 20mg Zocor is priced the same as an 80 mg tablet. 75% savings can be achieved if an 80 mg tablet is cut into 4 pieces. Dr. Houck's favorite Statin is now Vytorin which is a combination of Zetia and Zocor. It is priced less than Zocor and is two medications. Zetia blocks cholesterol uptake within the gut and reduces dietary cholesterol exposure. It comes as 10 mg tablet or in combination with Zocor and therapeutic efficacy show that 5mg works as well as 10 mg. Splitting this pill is then the cheapest and most effective drug for lowering both cholesterol and inflammation. The price advantage of pill splitting in a typical diabetic is now demonstrated in the following example. The first scenario is without pill splitting and the second is with pill splitting demonstrating a savings of nearly $52,000 over 25 years. Since most of our population is becoming diabetic these numbers apply to most of the population. If we would apply this to 20% of our current population 40,000,000, which represents the diagnosed and un-diagnosed diabetics, it would save $2,080,000,000,000,000. (2 Trillion dollars) The medications listed are really the bare necessities for a diabetic patient.

Treatment of:	Drug	Script	Cost
Platelets	Aspirin 325mg/ Plavix	I PO q d	5/100
Hyperlipidemia	Zocor 20mg	I PO q P.M.	$116
	Gemfibrozil 600mg	I PO q A.M.	$8
Hypertension	Hctz 25mg	I PO q A.M.	$8
	Lisinopril 20mg	I PO q A.M.	$62
	ToprolXL 50 mg	I PO q A.M.	$20
Glucose	Metformin SR 500 mg	II PO q P.M.	$80
	Actos 15 mg	I PO q d	$102
	Insulin NPH/R		$35/35
			$471/$519
One Year			$5652
25 Years			**$141300**

With pill splitting

Treatment of:	Drug	Script	Cost
Platelets	Aspirin 325mg/ Plavix	I PO q d	5/100
Hyperlipidemia	Zocor 1/4 of 80mg	I PO q P.M.	$29
	Gemfibrozil 600mg	I PO q A.M.	$8
Hypertension	Hctz 1/2 of 50mg	I PO q A.M.	$4
	Lisinopril 1/2 of 40mg	I PO q A.M.	$23.50
	Metoprolol 50mg	I PO bid	$9.50
Glucose	Metformin 500 mg	I PO bid	$46.75
	Actos 30 mg 1/2	I PO q d	$75
	Insulin NPH/R		$35/35
			$270.75/370.75
One Year			$3249
25 Years			**$81225**

Houck saves per patient over 25 years   **$60075**

The next three medications for review are all thinner agents that make blood less likely to form clots. Aspirin and Plavix are anti-platelet medication. Warfarin is an anti-thrombotic. As a review from chapter two, the platelets are small elements in the blood that attach to abnormal endothelium and initiate clotting by clumping together. Aspirin inhibits the platelets thru inhibition of Thromboxane A2. Plavix blocks the effect of ADP. Coumadin (warfarin) is an anti vitamin K medication that interferes with coagulation.

Aspirin

Felix Hoffman invented aspirin in 1897. His goal was to relieve his father's pain from arthritis. Salicin is a component of willow bark used for pain relief for centuries. Hippocrates himself may have used this early form of Salicylates. Salicylates were difficult to tolerate due to stomach upset, and by adding acetyl to the chemical formula the drug had fewer side effects. Aspirin is now in many products and has been over the counter since before the days of the FDA.

It is unlikely it would be approved if it were introduced today. The high rate of GI bleeding and Reye's syndrome (a deadly childhood disease associated with chickenpox and the use of aspirin) would have doomed this medication for public use. The drug is amazing in that it was the first medication to demonstrate a reduction in death from myocardial infarction (MI). When compared to potent thrombolytics medication it was nearly equal to the clot busting capability and was found to enhance the effect of these medications. Now it is standard procedure for patients to take an aspirin as soon as chest pain occurs and indeed this has aborted some myocardial infarctions. In the ER, patients are asked to chew the aspirin so it can be absorbed quickly. Aspirin affects the platelets by inhibiting them from sticking together. Aspirin in small doses can be effective. In small doses, aspirin can be effective. As small a dose as simply licking a pill might be enough to poison your platelets. The downside is the blue tattoos seen on old hands and arms that demonstrate skin hemorrhage from minor trauma. I don't recall seeing anyone with a heart attack who presented with the biologic evidence that the aspirin was working (those with blue-black skin). Aspirin is not proven to be as effective in young women as compared to women aged 65 and above. (The Women's Health Study)

Plavix

Clopidigril is a "super aspirin" that inhibits platelet function by inhibiting ADP. It has allowed the development of stents, which are Chinese finger traps that can be flattened onto balloons, and blown up in arteries. The artery is held open by the small metal struts of the Chinese finger traps. At the end of the 1970 decade Major (Dr.) Schatz of the Army and Dr. Palmero, a radiologist at University of Texas in San Antonio began putting these hand made devices into pigs, baboons, and rabbits. The body rejects anything foreign, and blood will clot on anything that is not endothelium. These animals and patients needed intense anticoagulation to prevent a blood clot from forming on the stent. In several weeks the endothelium would grow over the stent and the chance for thrombosis would be reduced.

A billion dollar industry was born. Stent placement received a big boost when it was demonstrated that the combination of aspirin and Plavix would prevent stent thrombosis. Re-infarction and stroke have been reduced by Plavix. It makes one consider, is the stent necessary or is all of the benefit from the drugs given with the stent?

While Plavix contributes to the prevention of stent thrombosis, it also has negative effects. For instance, if the patient bleeds from Plavix, a large quantity of blood is lost and all that can be done is to wait for the drug to be excreted from the system and replace the blood. Surgeons are excessively concerned with bleeding and express a great dislike of this medication. They want Plavix to be stopped 5 days before an elective procedure is performed. The only physician who should stop this medication is the cardiologist who placed the stent. Plavix can not be stopped until healing is assured because stopping the Plavix suddenly can cause an immediate heart attack. Bare metal stents require one month of Plavix and drug eluding stents are now recommended to have Plavix continued for one year.

Judy is a nurse who worked too hard. She had performed many tasks as a nurse and had slowly been pushed into the clipboard business of nursing because of her people skills. Physicians, nurses, and the administration respected Judy for her knowledge. The administration soon realized that if there was an emergency, a short deadline, or a regulatory problem she was the go-to person because she got things accomplished. The frequent 12-hour days and weekend reports made her lifestyle a great example of poor health. She had type II diabetes and frequent urinary tract infections. She came to cardiology when walking across the parking lot caused tightness in the chest and breathlessness.

She had a cardiac cath that demonstrated diffuse disease in all three vessels with the target vessels being poor quality. The Houckster decided to try medications since her heart was functioning at normal levels. She was loaded up on beta-blockers, aspirin, Statins, ACE inhibitors, and nitrates. She walked on the treadmill for 10 minutes before getting her angina. She did well for 6 years.

On a routine visit her LDL Cholesterol was 65, but her High Sensitivity C Reactive Protein (HS-CRP) was 6 giving her an increased risk of heart attack or stroke. In his note, the Houckster suggested that Plavix may be indicated, however, her angina had not changed and there was no data to support his gut feeling, that she should be on a $4 per day medication.

Within a month the little voice in the Houckster's head was now screaming, "I told you so." Judy awoke with intermittent chest pain one night. Typical of females even with her medical training, she did not want to be a bother in the middle of the night. Although she knew better, she waited until the morning before calling the cardiology clinic. Soon after she arrived in the clinic she was taken to the cath lab. All of her vessels improved in caliber and she showed regression on her medical therapy. Unfortunately, she now had a clot in the improved circumflex. A stent to this vessel removed the problem and assured she would now be on Plavix. The Houckster would keep her on Plavix for life. Judy continued to give more than the job required by beginning a daily exercise program and a reduction in her inflammatory markers.

Coumadin, Warfarin

Rat poison kills rats by letting them eat as much as they want so that their clotting factors fail and they bleed to death. Warfarin is rat poison We treat patients with a monitored dose of Warfarin so the bleeding time is prolonged only about twice as long. Warfarin has a low therapeutic index, which means that the difference between help and harm is narrow. Coumadin Clinics are the best means of accomplishing this task. Algorithms are developed and patients are tracked and sent reminders when they fail to show for their appointment. The patients are taught to be responsible and if they begin a new medication they need to have their blood clotting (INR) checked to see if there is an interaction. Any drug that displaces Warfarin from proteins will change the drugs effectiveness.

Warfarin is an anti-vitamin K medication. Any change in diet, such as replacing a salad diet with a high protein diet, will change the drugs effectiveness. Dr. Houck tells his patients to continue eating greens, just eat the same amount every day.

Warfarin can reduce the chance of having a stroke from atrial fibrillation from 5/100/year to 1/100/year. (Fuster et al, ACC/AHA/ESC Guidelines for the Management of Patients with Atrial Fibrillation, JACC Vol.38, August 2001:1231-65). It improves outcome after myocardial infarction. It is necessary after a pulmonary embolus. The drug can be administered safely with improved patient survival as long as surveillance of the drug is performed. Patients who are not candidates are those who are non-compliant, and people who are prone to falls. A wheelchair and a walker are not necessarily a contraindication to the drug, but a discussion with the patient and their family is necessary.

Leona was an 81 year old widowed housewife who lived in an old railroad mansion. She had heart failure and was well compensated. Her left atrium increased in size to 6 cm and she had been in atrial fibrillation for one year after having intermittent atrial fibrillation the preceding 5 years. She initially had a transient ischemic attack (a stroke that was reversed by the clot spontaneously breaking apart), but since starting Coumadin she had preserved all of her wits. Her clever wit made her visits a joy for the staff. She did not act her age and was proud of it. Her routine included a beauty parlor visit every Thursday at 9:30, so that her grey-blue hair would have the correct height and shape for the weekend.

Texas is a sunshine state with mild winters. South Texans are not comfortable with ice and snow. A mere dusting will result in hundreds of fender benders. Wednesday night the temperature dropped to 25 ^{0}F and the rain gave a fairy like appearance to the live oaks that were now covered in ice. The pavement and walkways were black with the "black ice." The quarter inch of ice made the roadways appear to be clear of ice, only darker in appearance. Leona was well aware the ice would cover the roads for hours, and was determined not to let it ruin her plans for the weekend. Despite listening to pleas from her family and without a doctor's permission she preceded to the beauty parlor. She called first to see if her girl stylist was there and it was a pretty sure thing she would be there, since the parlor was in her house. Her girl stylist told her not to come. Leona ignored her advice and still made an attempt to get beautiful. She was cautious but the black ice was unforgiving. She fell and hit her head accumulating a large amount of blood between the brain and dura mater the protective membrane surrounding the brain. The fall was serious, but Warfarin caused the fall to be fatal.

****

Murray returned to the heart failure clinic and greeted Tracie with a big smile and a hug. It was now 4 months since his first visit to the heart failure clinic and he had made remarkable recovery. Tracie who had seen many miracles was impressed with the 82 year old. He was on all of his medications. As stubborn and recalcitrant a patient could be in the beginning, Murray had become a model patient. He had not required Lasix (diuretic) for 3 months and was asymptomatic. He was successfully cardioverted from his atrial fibrillation and had received a $50,000 bi-ventricular pacemaker. His medications now consisted of Coreg (Beta blocker) 25 mg twice daily, Spironolactone (potassium sparing diuretic and anti-fibrotic)25 mg daily, Lisinopril (ACE inhibitor dilator) 20 mg twice daily, Digoxin (inotropic medication) 0.125 mg daily, Amiodarone (anti-arrhythmic) recently reduced to ½ of a 200 mg tablet, and Vytorin (Statin) 80/10 ½ tablet. He was still on Warfarin (thinner) but it would be discontinued by the end of the day. Tracie had ordered an echocardiogram, a BNP, renal function, thyroid and liver test.

These were monitoring tests for potential toxicity from his Amiodarone and to monitor his heart failure.

Murray knew he was better. He had graduated cardiac rehabilitation and was now training for his planned adventure. Tracie examined his lab and broke into a toothy smile and then a giggle. This was the favorite part of her job. The BNP was less than 20. The echocardiogram was normal with an ejection fraction of 60%. His entire lab was perfect. Tracie gave him the good news and immediately told him he had to keep taking his medication pointing her slender finger in a very menacing manner. Murray countered with a calendar with two weeks circled in January. He said she had to arrange for vacation because in 6 months he would have completed his training and they were going on safari in the Philippines during the January dry season. Murray arose from his chair and simply stated he had training and left in a hurry after a goodbye hug. He was now riding his stationary bicycle one hour a day walking 30 minutes and had hired a trainer to help develop stronger muscles. Murray had lost 15 pounds and planned to turn the next 10 pounds into muscle. If he was taking these young healthcare workers into the mountains in a tropical country he needed to be able to manage himself and them.

There are many other useful medications recommended to patients. Direct to consumer advertising and information packets can be misleading to patients. The number of symptoms that patients experience is much less than on the insert. The symptoms reported in the insert are from the clinical trials of the medications. Both the placebo group and the treatment group indicate the symptoms they are experiencing. Everyone has a headache, stomach upset and difficulty with sleep. The safety committee attempts to determine if these side affects are greater in the treatment group as compared to the control group. For example if impotence occurred in 4% of the treatment group and 1% in the placebo group, impotence would be listed as a side affect. The chance of having this side affect is however very small 4 out of a 100. If I tell the patient he may have impotence from this medication there is a tacebo (self-fulfilling prophecy) effect and 25% of those individuals will be too soft to perform. Taking medication is important. In trials the individuals who took placebos did better than the non-compliant patient taking the same placebo. Taking medications and a desire to get better is powerful in obtaining cure. One final medication will be described in this chapter. Other medications may have short descriptions in the glossary.

Nitroglycerine

Nitroglycerine is a medication that has had a curious development. It certainly has destructive power and has been used for mining, wars, and for relief of angina. Patients are concerned if they allowed their nitro pill bottle to fall onto the floor it would result in a deadly explosion. The explosion will not occur because the concentration of nitroglycerine is too weak to react. The pill is only a soluble container for the nitro that exists in a gas form. The pills need to be kept sealed in their dark container to prevent the gas from escaping. Old pill bottles will have the pills left intact but the nitro will have evaporated into the atmosphere. The unsuspecting patient will not note a slight burning under the tongue or the characteristic flush and headache when they put the pill under the tongue. They will not have their chest pain relieved either because there is no medication in the empty pill. The gas will quickly dissolve into the venous plexus under the tongue and travel to the heart and brain in seconds. Relief of angina is accomplished by decreasing preload. The reduction of preload decreases the size of the heart. A smaller heart has less oxygen consumption relieving angina. The dose that dilates coronary arteries is not known. It is likely 400 micrograms and this is the size of the Nitroglycerine tablet.

The affects of nitroglycerine became evident in munitions workers. During the week they were exposed to the gas and had vasodilatation of their coronaries. On Sunday or Monday morning they would have chest pain because of withdrawal. The medicinal therapy with nitroglycerine began with this work related illness.

The use of nitroglycerine in the treatment of angina is not controversial. My staff doctor chastised me when I failed to give the sublingual form to patients who were discharged after a myocardial infarction. Was this berating deserved? Nitroglycerine has never improved mortality or reduced infarct size. In fact, some EMS systems have removed the drug from the ambulance because it was thought to be too dangerous.

In acute myocardial infarction due to blood vessel closure it is unlikely that Nitroglycerine will help and in some cases such as a right ventricular infarct it will precipitate profound hypotension that could result in death. My own feeling is in conflict with the guidelines and I feel that sublingual Nitroglycerine is contraindicated in acute myocardial infarction since it has no benefit and potential harm. Intravenous Nitroglycerine can be given in a safe manner because it has a short half-life. Usually the emergency room starts at 10 micrograms per minute and does not bother to titrate.
It is unlikely that this small dose has any benefit other than making the doctors and staff feels like they have accomplished a goal.

The proper use of Nitroglycerine is for angina. The patient should stop their activity that precipitated the chest discomfort and sit down. They should not lie down because this will increase preload and worsen the pain. The small tablet should be placed under the tongue while seated. You should never be standing or a pass out spell from a reflex bradycardia may occur. If pain is still present after 5 minutes EMS should be called to haul you to the nearest emergency room.

Nitroglycerine can also be given at bedtime for congestive symptoms of heart failure, or for acute shortness of breath from pulmonary edema. If a patient knows an activity will cause angina, they can take sublingual Nitroglycerine before they begin the walk to the mailbox or their morning exercise routine.

Nitroglycerine has been the go-to drug for angina. As shown above it has great use in angina and heart failure, but it can be misused with potential disaster. Tom Brady was on the eighteenth hole and was one putt from winning the pot. He had never been in this position and tension was mounting. He noted a little tachycardia and slight chest discomfort. He did not want to draw attention to himself so he casually slipped a Nitro under his tongue. There was slight burning as the gas entered his system and dilated his dehydrated constricted veins. The heart suddenly emptied as the blood remained in the now dilated veins in the legs. In an effort for the heart to save itself from an empty state it slowed to give more time to fill the empty chambers. Tom collapsed and EMS was called for what was thought to be a heart attack. He awoke quickly when he hit the ground as the preload was corrected by the supine position. EMS arrived and he had to forfeit the match.

Sally J. was a 23 year old getting ready for her first date when she swallowed a hard candy. She had onset of chest pain. She went to the emergency room where she was triaged to chest pain despite her young age. Forty seconds after the nurse placed sublingual Nitroglycerine under her tongue she became hypotensive and unresponsive causing terror in the ER. She was admitted and missed her first date. She had a Bezold-Jarisch reflex from the nitroglycerine.

Mrs. Peabody was a 72 year-old female who developed chest pain associated with nausea, vomiting and extreme weakness. She was placed on the monitor and was bradycardic at 40 beats per minute with a blood pressure of 100. Sublingual nitroglycerine was given. Asystole was noted on the monitor and an unsuccessful code ensued.

Samuel G. had used Viagra prior to his out of marriage affair. He failed to tell the emergency room doctor that he had taken this medication when he arrived with chest pain. The nitroglycerine caused hypotension that required 4 liters of fluid and Epinephrine to maintain blood pressure. His hospital stay was extended 4 days because of his shock at presentation. It is very important to disclose to your physicians all the medication you are taking.

Nitroglycerine can relieve pain but in some circumstances can kill or harm.

Ranolazene

The last medication to be reviewed is the only anti-anginal medication developed in the last 25 years. As with many drugs the mechanism of action during development is not the way the drug acts. It is thought to inhibit the late sodium channel currents and by this means prevents the build up of calcium in the cell during diastole. Calcium build up occurs during ischemia and is manifested by an S4 and diastolic dysfunction. The drug should be very good for diastolic heart failure. The drug prolongs the QT interval so there was concern it may contribute to arrhythmias. In fact arrhythmias are reduced. The drug has great promise for individuals who have severe diastolic heart failure and recalcitrant angina. The dose is up to 1000 mg two times per day but many times in association with renal failure 500 mg at night is enough. The medication does not change the heart rate or blood pressure so it can be used in the difficult patient who has to little of both. For some unknown reason it improves diabetic control by lowering HgbA1c. The drug makes patients feel better. It has not been proven to make them live longer.

This chapter reviewed a few common medications used in the treatment of heart failure. Medications will work if appropriate doses are faithfully taken by patients. The art of medicine is getting your patients to comply. This chapter did not cover drug-drug interactions and this has to be considered whenever a new agent is prescribed. Epocrates.com is a good source for learning about drugs and interactions.

****

Murray's brain was aching from all the information Tracey had pounded into his brain. He knew he would remember very little of her lectures, but he understood the principles. If his heart was to get better, his body had to be a better environment for his heart to function. That environment meant low blood pressure, low heart rate, and allowing his body to adjust to the medications that would repair his damaged heart. The woozy feeling he had when he would stand up was improving and the fatigue was slowly subsiding. The changes were slow but he was thankful that he was not drowning. He was impatient to get better and vowed he would obey that shagged headed nurse practitioner.

With a sudden leap to his feet, Murray went for the door a second too late. Tracey handed him a consult to the dietician.

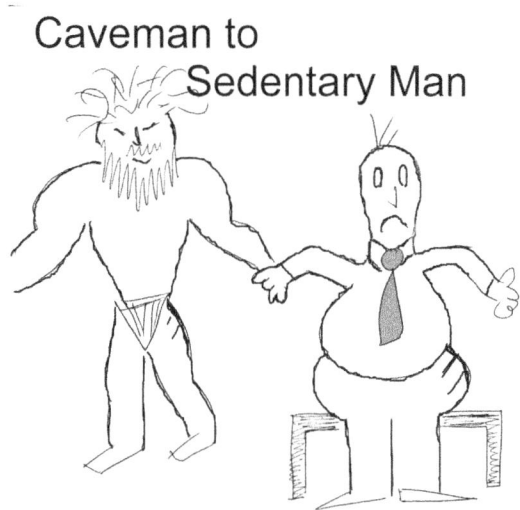

700 Pound Man

The 700 pound 32 year old male was in beds 7 and 8 of the intensive care unit. The beds in room 7 and 8 had been bolted together by maintenance. The curtain between the rooms 7 and 8 of the hospital bay was removed. His weight increase of 200 pounds in the last 2 years was too much for his heart to maintain. A simple case of bronchitis resulted in respiratory failure. The diaphragm, which works like a bellows to expand the lung, has to lift 100 pounds or more with each breath. As a result of this great work the diaphragm adapts a strategy of resting between breaths. The effect of this rest drops blood oxygenation levels causing the pressure in the lungs to rise to very high levels. The right cardiac chambers that have to pump blood to the lungs begin to fail with a back up of fluid into the tissues. The most recent weight gain was due to 50 pounds of fluid retention. The patient could no longer perform the work of breathing and was now dependent on a ventilator to perform this lifesaving task.

The surgeons who could perform a gastric bypass felt they could not do this procedure until he dropped below 500 pounds. His hospitalization would be costly and would likely result in back-injured employees who would use super human strength to rotate the patient to prevent skin breakdown. Soiling the bed and the required bed change would be a disagreeable and daunting task. New hospitals are being built with industrial cranes to make this process easier.

Despite the vast number of stored calories, nutrition would be provided through a nasal gastric tube. The excess weight would be reduced in a slow and deliberate manner to help prevent the loss of muscle mass. Once 500 pounds was achieved, the gastric bypass would help prevent overindulgence and cause satiety with fewer calories.

290 Pound Female with chest Pain

The obesity epidemic was also affecting the cath lab requiring stronger tables and greater expertise in vascular access. The 5 foot tall, 290 pound female was wheeled into the cardiac catheterization laboratory. Her complaint was progressive shortness of breath and fatigue with some chest heaviness. The patient used to be a cheerleader in high school; she was married within one year of high school and had four children in succession. She had been working as a receptionist for the last 15 years and then had a back injury after a fall on a wet floor. She became disabled due to constant back pain and the inability to sit in a comfortable position. Consequently, Louise gained nearly 75 pounds in the last year and a half.

Several folds of fat covered the inguinal area that was scaly red. The pulse was quite difficult to feel and the cardiology fellow would apologize many times as he attempted to place a needle into the femoral artery that was hidden somewhere in mounds of fat. She had arrived in the cath lab because her non-invasive studies were equivocal. A Dobutamine Echocardiogram was unsuccessful due to poor images. Her thallium demonstrated an anterior defect that was thought to represent ischemia but shifting breast artifact could not be ruled out. Due to her back injury, a functional assessment by a simple ECG monitored treadmill was not obtainable.

Luckily, the coronary angiogram demonstrated only luminal irregularities with no significant proximal disease detected. The patient required only medications and did not need surgery or angioplasty. She was lucky because both procedures had greater risk when weight was greater than 175 pounds over ideal.

300 Pound Software Engineer First Risk Factor – Low Birth Weight
The second patient on the Lab list was a 300 pound six foot tall computer software engineer. He had never been athletic and tended to play with video games ultimately leading to his current career. His large stature seems somewhat unusual due to his pre-maturity and birth weight of only 4 pounds. He had one episode of severe chest pain that was substernal and associated with diaphoresis. He had some nonspecific electrocardiogram changes consistent with ischemia or hypertension. For this reason he was sent directly to the cath lab to determine his anatomy to see if a myocardial infarction was in his immediate future.

Cardiac catheterization tables 10 years ago could only hold 250 pounds. If the patient was more than 250 pounds the procedure was felt to be unsafe because of potential breaking of the table with a patient falling on their head. Lawsuits discriminating against obese individuals have changed this attitude. Regardless of the patient's weight, cardiac catheterization is offered if clinically indicated. Tables now can hold up to 350 pounds, and unfortunately, there are some patients who exceed this weight limit.

During the last twenty years the weight gain has accelerated in the population. Although the previous mentioned encounters are extreme examples, they reflect the United States, and the rest of the world, seeing an obesity epidemic. Airplane and sports seats have gotten larger over the last century. Kind passengers are buying two seats to make their travel plans, while others simply overflow into your seat. There are now more obese people than those who are starving. It is estimated that there is 1 billion overweight individuals and 300 million obese. Twenty two million children under the age of 5 are overweight. (World Health Organization. Obesity: preventing and managing the global epidemic. Report of a WHO Consultation presented at: the World Health Organization; June 3-5, 1997; Geneva, Switzerland. Publication WHO/NUT/NCD/98.1.

Calorie
A calorie is a unit of energy. For every 3,500 calorie intake over output, 1 pound will be gained. The corollary is that it takes 3,500 extra calories burned by exercise to decrease 1 pound. The recommended weight-loss program would reduce 500 calories per day. The 3,500 calorie reduction over one week would result in one pound weight loss. To lose 50 pounds at this rate would take one year. A more reasonable goal would be a 10 pound weight loss that would take five months by decreasing 250 calories per day. However, dieters are often inpatient and have no intention of waiting five months to lose 10 pounds. Initiation of the diet and the expected result are disconnected and the results lag many months behind the initiation of the diet. The slow pace discourages most dieters and allows them to fall back into their previous bad habits well before their goals are achieved. In dieting there are as many errors as there are diet books.

For the most part, diets are ineffective in maintaining weight loss. I do not recommend any diet. The necessary elements in a successful weight loss program have to revolve around several principles. The first principle involves a daily measurement of caloric intake. Simple avoidance of high calorie snacks may be all that is necessary to halt a steady increase of weight gain. The second principle is an exercise program with resultant calorie output on a daily process. The basal calorie intake is different for each individual. It depends in part upon the patient's efficiency of metabolism. The metabolism can switch gears if it senses starvation. Metabolism is dependent on the number of hours of wakefulness as well as the activity level while one is awake. The average calorie intake is approximately 2,400 calories. Metabolism varies between individuals and can vary within the individual.

The goals of well-balanced nutrition should be reinforced on a daily basis with a measure of calorie intake versus calorie outtake. One day per week can be taken as a free day where calories are not measured. There are several very good programs that can help individuals with this process such as *Weight Watchers* and *Balance_log* which is a program that can be placed into a Palm Pilot, or smart phone. The food selected and serving size is entered. The amount of exercise performed and intake of food calories are all placed within a daily database. The program will then calculate the number of calories reduced over the basal metabolic rate. The basal metabolic rate can be estimated through the program or actual determinations can be made through a measure of oxygen consumption. Daily feedback is then the method of goal setting. This technique takes discipline and both the calories and exercise performed must be placed into the database. The program is friendly and easy to use but still requires discipline, input of the data, and even more discipline to avoid extra calories once the program has told the operator that the daily required calories have been met.

Strawberries will be used as an example to understand this concept. Strawberries could be an excellent pick up line. One strawberry is 4 calories and 16 strawberries would be 64 calories. The amount of energy released by kissing and hugging for 53 minutes, or moderate sex for 41 minutes, or vigorous sex for 35 minutes is equivalent to 64 calories – 16 strawberries. You could suggest sharing 8 of the 16 strawberries with your partner. By only eating 8 strawberries you will consume 32 calories and would expend 64 calories for a net loss of 32 calories the equivalent weight loss of .15 ounces. Making love 1 hour per day for a year and eating 8 strawberries each time would cause a total weight loss at one year of 5.87 pounds. If you skipped the strawberries so you had no calories intake and made love for one hour per day, you would lose 11.73 pounds.

The strawberry example illustrates the concepts of conservation of energy. After using this example in lectures and sidewalk advice, I have been getting anonymous strawberries from everyone. If I was single I would always carry strawberries as a type of advertisement. Any discussion of chocolate covered strawberries is beyond the scope of this book.

Successful dieters will lose weight: as they restrict their calories, they will become smaller. Unfortunately, at a smaller size their body requires fewer calories and to lose additional weight more calories need to be restricted. Diet is a difficult four-letter word. Weight loss is a slow process requiring patients to think about everything that goes into their mouth.

Mystery of Obesity
Why have the people of the world grown larger? We can start by looking back at our cavemen ancestors. In their world, calories were not in abundance. For the caveman it was quite beneficial to store calories. It allowed them to survive during wintertime when food was quite scarce. The excess calories were likely stored in the fat so it could be readily exchanged as an energy source when food was no longer available. The fat was initially stored outside of the abdominal cavity. If the abundance of the food supply were very great the visceral fat within the abdomen would then become the target fat cells for the extra availability of food. The visceral fat would be the first to be depleted when food became short in supply. Visceral fat was never meant to be a long-term fat storage for more than one season. The caveman had an attitude of "see food - eat food.". Even if the caveman happens to come across a decaying carcass, the Leptin hormone within their gastric mucosa would trick the brain into believing this was a delicacy and not a disgusting mess. In some individuals the effects of these hormones can be quite exaggerated and results in continued weight gain similar to our 700 pound young man. For most of the population, however, the "see food - eat food" survival technique is the reason for the dramatic increase in body size. Fat deposits have increased as food became more plentiful and more calorie rich. Unfortunately, with calorie excess, our visceral fat tends to be the target for these excess calories. Our abdominal girth begins to increase in size. The point of no return is 40 inches in a male and 35 inches and a female. (Ford et al, Prevalence of the Metabolic Syndrome Among US Adults, JAMA, Vol. 287 No. 3, January 16, 2002) As the abdominal girth grows beyond these parameters the excess storage of fat becomes a metabolic time bomb.

The reduction in daily activity only compounds the problem of increased caloric intake. Our great-grandfather's work in the fields expended many calories to produce food. Currently, even our high work level farmers of today have reduced calorie output as they drive around in their air-conditioned $200,000 tractor with DVD player and global positioning sensor. In today's world we sit in front of computer screens, televisions, and expend very little calories while we are ingesting even more calories. The epidemic of obesity is also seen within our children for the same reasons. Unfortunately, the effects of obesity in our children may have long lasting life consequences. Children are now developing Diabetes Mellitus type II by the age of nine and will likely have coronary artery disease and its consequences in their early 30s. Our current downtrend in cardiovascular deaths could rapidly swing upwards due to the great burden of diabetes that is forecasted.

This trend of increasing obesity and increasing diabetes has been plotted for the last 25 years and is rising at an alarming rate. The economic impact of this epidemic can hardly be estimated. The number of sick days per month for the working population will increase. Medical retirements due to myocardial infarction, stroke, and peripheral vascular disease will result in the population having a younger retirement. The medication expenses to maintain an individual with metabolic syndrome currently costs $600 per month (see chapter 4), the equivalent of putting a child through college every 10 years. Hospitals will be reimbursed less and less per patient because of the burgeoning costs. This will drive down salaries and medicine will no longer attract the brightest and best of our student population. Legislation will do nothing to swing the epidemic until the individual patient has incentive to change their lifestyle.

The progression to diabetes from obesity is a complicated pathway and is certainly not understood by this cardiologist. In simplest terms, for type II Diabetes, the body outgrows the size of the pancreas; and diabetes ensues simply because the pancreas cannot produce the amount of insulin required by the larger body. The good news is that diabetic patient can become non-diabetic by simply shrinking in body size so that the body again becomes a match for the pancreas. This simple explanation for diabetes is quite inaccurate. In a diabetic patient, the insulin levels climb to alarming levels before the glucose levels rise. The tissues of the body become less sensitive to the effects of insulin. The pancreas attempts to initially compensate by producing more insulin. The stimulus for increased insulin may be simple increase in glucose levels but there are other factors involved. Eventually the insulin levels begin to drop with the burnout of the islet cells and patients have to go on injections.

Inflammation
One of the new popular theories that explain many of the multiple disease processes associated with obesity and diabetes is inflammation. A diabetic patient demonstrates greater levels of inflammatory markers than the non-diabetic patient. Vascular disease patients have more inflammation than patients without vascular disease. Degenerative joint disease with resultant inactivity leads to both obesity due to less caloric output,

diabetes due to weight gain, and ultimately to death from a stroke or myocardial infarction from inflamed blood vessels.

Is it simple body size that causes knees and hips to wear out? Most patients entering the office to obtain a pre-operative evaluation for knee replacements were obese. It made sense that the excess weight on the joint would cause it to wear out more quickly. This premise is incorrect and body weight has nothing to do with the destruction of the joint. After all, elephants have good joints. The inflammatory state that caused the obesity and caused the diabetes is what is currently causing the knee arthritis. These same patients will likely suffer myocardial infarctions because of the increased inflammation in their blood vessels causing their atherosclerotic plaques to become thin, rupture, or erode. The thread that holds the connection between obesity, diabetes and vascular disease is called inflammation. Perhaps a new theory of obesity implicating inflammation should be written. The new diet book would be called *Obesity and Inflammation a Deadly Companion*. The book would become quite popular and outsell the Atkins diet, the South Beach diet, and would likely come in as a tie for the Mediterranean diet. The author would then be rich, famous, wearing a funny jogging suit and would not have to complete writing this book.

The thread between these disease processes is inflammation. The most inflammatory process for humans occurs when we eat. The gastrointestinal tract does not simply process food but has a major function in our immune system. During eating we take in many foreign proteins that are to be broken down in the gastrointestinal tract, however, some of these proteins are absorbed directly into our lymphatic system and into the bloodstream. Our immune system has to decide if these are friendly or foreign proteins. These foreign proteins stimulate the inflammatory process. The inflammatory process disconnects our body weight maintenance program and results in obesity. Inflammation within the islet cells of the pancreas and in the fat cells of our visceral fat behaves badly. The inflammatory process spills over into blood vessels and causes coronary atherosclerosis. The inflammatory process targets the joints as simply another tissue to torture. Joint pain is a result of our poor eating habits

The major flaw in my plans to write a dietary advice book is that I do not know which foods cause a reduction in inflammation. I do have suspicions that animal fat with multiple proteins attached, taken in excess is guilty of raising inflammation. Most fruits, vegetables and nuts, and monounsaturated fats such as olive oil have little to do with inflammation and may actually lower our inflammatory state. The next element in decreasing inflammation is simply bulk and fibre that will speed the transit of the abnormal proteins through the digestive system so they are not absorbed but excreted. Human defecation should look like bear scat.

My diet would likely be very unpopular because it would have the side effect of large smelly stools that could plug up the plumbing. The other aspect in elimination process is exercise. The biggest human change over the last 10,000 years is the reduction of calorie output in this simple form of exercise. The automobile, the grocery store, home delivery, elevators, escalators and bicycles with so many gears that peddling is effortless, have contributed to our calorie deficits. If we preformed an adequate amount of exercise most of us would likely be able to eat what they desired. In our current exercise-poor environment we have to limit the amount of enjoyable food that we partake. Exercise speeds the transit of processed food in the gastrointestinal tract again lowering inflammation. When studied, exercise continuously shows beneficial outcomes in regards to fewer adverse cardiac events. A government website sited is useful. http://www.cdc.gov/nccdphp/dnpa/physical/importance/why.htm. It is difficult to show a change in the function of the heart as a result of an exercise program. There are other aspects of exercise on the heart other than conditioning. Studies have demonstrated a benefit of exercise by a reduction of inflammation. Weekend warriors typically experience an effect contrary to a routine athlete. In the case of the weekend warrior, there is an increase of inflammation due to the wear and tear of the muscles. On the other hand, an individual in a committed exercise routine will have a reduction in inflammation. The soreness experienced by weekend athletes can be attributed to inflammation. Starting an exercise program can be dangerous because it increases inflammation. A gradual increase in exercise can be accomplished quite safely. The danger is not starting an exercise program but stopping and restarting. Exercise needs to be a life long daily commitment.

The fattening of America seems to have begun when both mother and father began working outside of the home. Home cooked meals are a rare event instead of a nightly ritual. The fast food industry quickly filled this necessary gap. The size of the meals has increased exponentially and can be tracked with the increasing weight gain of the population. Soft drinks such as Coca-Cola can be used as a historical tracking method for portion sizes. The 7 ounce Coke in a glass bottle was very satisfying. The typical serving size now is more than three times that amount. The food that is served with this soft drink is not only great in quantity, but also excessive in calories. Potatoes soaked in oil, fried, and covered in salt is about as close to a coffin nail as a cigarette.

Houckster-size
While seated at your favorite restaurant please request to have your meal Houckster-sized. A Houckster-sized meal is one that is divided into two portions before it is ever brought to the table. Half the portion is placed in a box to be taken home and consumed in the next meal or two. The other half of the portion is brought to the table and enjoyed in the restaurant. This request for a Houckster-sized meal removes the primal reflex "see food - eat food." Why do restaurants serve such large portions? The cheapest commodity in the restaurant is the food. Proprietors of restaurants want to be sure their patrons are well fed and if they sent one customer out hungry it would be a violation of unwritten code of restaurants. People tend to flock towards bargains and a bigger size is a deadly bargain.

Identifying the above food traps is sufficient for most of us. If we are aware of these traps we can avoid obesity. For 5% of the population, and most of the morbidly obese, there is a genetic predisposition. (R. J. F. Loos, C. Bouchard (2003), Obesity is it a genetic disorder? Journal of Internal Medicine 254 (5), 401–425.) The genetic predisposition to store fat may have been of great benefit to survival during the times of the caveman; but, it is now decreasing lifespan. There may be hope as we unravel some of the reasons for the insatiable appetite that some people have developed.

Rimonabant is an endocannabinoid receptor blocker that may soon be approved for smoking cessation and weight control. The drug was originally developed for those addicted to marijuana. The connection to weight loss is that marijuana gave people the munchies and blocking the receptor could block the munchies. Like all drugs, Rimonabant has side effects including cessation from smoking and weight loss. Since there are many more people who suffer from obesity than marijuana addiction, the development of the drug changed. It has been studied in clinical trials and appears to be safe except for some rare suicidal ideation. (M. Randall, D. Kendall, A. Bennett, S. O'Sullivan, Rimonabant in obese patients with type 2 diabetes. The Lancet, Volume 369, Issue 9561, Pages 555-555)

As long as one is taking the drug, weight remains stable, but if the drug is stopped, the weight gains returns. Rimonabant would have to be given for life to maintain the weight loss. A drug given to cure addiction would become an addiction to maintain weight. In addition to the weight loss there is some direct metabolic effects that improve cholesterol and diabetic levels. The medication appears to be too good to be true and it is likely a segment of the population will experience side effects such as depression from this medication. So far there has been no harm detected from the slight chance of depression, however, the FDA is requiring more evidence. The drug can be obtained in other parts of the world.

As a result from implications of previous weight loss medications, current products are under constant scrutiny. Previous significant side effects from now banned weight loss medications include damaged cardiac valves and rare pulmonary hypertension. Although there are a great number of lawsuits related to these medications, actual harm from these medications appears to be quite small. Eliminating these useful drugs could cause more deaths than those who receive the adverse response.

Leptin, discovered in the mid-1990s, is a protein hormone produced in fat cells. As the fat cells get larger, Leptin production increases. This hormone then informs the brain to decrease appetite. It is also produced by some of the cells in the stomach and may be responsible for feeling pleasantly full or still hungry after a reasonable meal. The receptors for the Leptin lie in the hypothalamus and regulate body weight, metabolism, temperature, and appetite. Mutations have been discovered where the protein is absent and this results in increased weight gain. There can also be mutations in the receptor site of the Leptin molecule that will prevent its normal function and also results in obesity. If one measures the concentration of Leptin in obese patients, the protein is generally elevated. The majority of patients with obesity do not lack Leptin, rather they lack a Leptin resistance to the effects of this protein, causing their body to ignore this signal to decrease their caloric intake. Leptin can be injected much like insulin to help reduce body fat. If Leptin is given to normal size individuals they will lose weight. Unfortunately, if Leptin is given to Leptin insensitive patients, it will not curve their appetite or improve their obesity. There are regulatory proteins produced in the stomach that tell the brain there has been enough calories consumed. Other proteins regulate the speed of emptying the stomach. These regulatory systems in some individuals become cross-wired so that they are constantly hungry and constantly looking for food to eat. We all likely have some abnormal trait that can rewire our fat regulation. It is this genetic wiring that causes obese individuals to remain obese. By changing this regulatory system, the appetite can be suppressed and weight loss can potentially occur without the pain of hunger.

Summary

In summary, if one consumes excess calories as compared to the amount of calories expended in exercise - weight will go up. The fat cells become unregulated with unsuppressed appetite. The lack of exercise and increase caloric intake will result in inflammation, more diabetes, heart disease, obesity and miserable lifestyles. Sheer will, caloric monitoring, and beginning a lifelong exercise program can combat this major risk factor to health. While a fat pill may be on the horizon, for most of us, lifestyle is the answer. Exercise is the answer.

The 700 pound man was successfully weaned off the ventilator but had to use a nightly CPAP machine to keep from developing respiratory failure at night. Through continuously measured calorie intake and slowly progressive physical therapy he was able to lose 200 pounds over the next nine months. He underwent gastric bypass surgery and subsequently lost an additional 200 pounds. Other surgeries were required to remove the excess skin; the pannus of abdominal tissue left over after the weight loss was successful. The now 300 pound man had a new life that incorporated the philosophy of "eat sensibly and exercise often."

Gladys, the dietician, spent more than hour with the crusty patient. He had a charm that could not be defined with the appearance on an 80 year old patient. What Murray had was a lust for life, a purpose that drove his bones forward. The positive outlook was infectious and would spread to support staff and schedulers. The staff would do anything for Murray. The skull on this particular individual was, however, unusually thick when it came to dietary management. The old veteran had never been overweight but had a lust for salt. Salt was needed for his eggs, cantaloupe, and every snack food in his pantry had enough salt to last a lifetime. Murray would throw away $150 of groceries to abide by Gladys's demands.

Murray thanked Gladys for taking his remaining vice and placing it on a shelf out of reach. He would include fibre, fruits and vegetables, nuts, and avoid saturated fats and fried foods. He was contemplating his visit to the library to research Mayan history when Tracey provided him with another consult. She simply said with an evil smile that he was well enough to begin an exercise program and flashed a consult to cardiac rehabilitation. For some unknown reason a vision flashed into Murray's brain of Tracey in black leather and whips.

Chapter 6
Exercise, Ponce de Leon, and Stem Cells

Prometheus brought knowledge and fire to the earth. This kind act to humans was rewarded by torture. Prometheus was chained to a rock and vultures would come and eat his liver during the day and during the night his liver would regenerate so the entire cycle of agony could begin again. We are indebted to Prometheus for his sacrifice to give us knowledge. Prometheus has also pointed the way to cellular therapy and rejuvenation of organ systems. I was taught that the heart was completely differentiated and the cells that were in the heart could not be replaced. The evidence for this statement is that after a myocardial infarction and cell death the heart wall was replaced with scar, fibrotic tissue, and not active muscle. The liver can regenerate itself when damaged; however, previous teaching indicated the heart does not have this capability. Should this teaching be accepted?

Can the heart survive for 100 years on its original cellular matrix? The heart contains 1.3 $x10^7$ cells. 95,000 cells die each day due to apoptosis and necrosis. (Piero Anversa, , Jan Kajstura. Ventricular Myocytes Are Not Terminally Differentiated in the Adult Mammalian Heart *Circulation Research.* 1998;83:1-14.) At this rate, the heart would run out of cells in five months. Anversa first challenged this hypothesis when he investigated cardiac hypertrophy. In patients who had pressure overload from valvular heart disease or hypertension, Anversa found that the cells not only were increased in size, but were also increased in numbers. By carefully counting the number of cells in a cross-section of the heart he was able to demonstrate that the number of total cells was increased. He realized the only way that this could happen is if there were new cells dividing and growing within the heart.

After many years of criticism and numerous attempts at special stains, small precursor stem cells have been identified within the heart to support his theory. Other evidence has come from patients who had transplanted hearts. Male patients who received female hearts after transplantation have undergone biopsy procedures checking their new hearts for rejection. In the biopsy specimen from the female donor heart, cardiac tissue with Y-chromosomes was identified. The Y chromosomes could only have come from the male recipient who no longer had his heart because it was removed during transplantation. These cells could only have come from circulating stem cells from the male. The female heart was being transformed into a male heart. This could only happen through a rejuvenation process.

Clinical trials are now underway to investigate the possibility of repairing a damaged heart after a myocardial infarction. Cells are being injected into the hearts circulation or directly into the muscle itself. There are different cells that are injected. Cells harvested from the bone marrow are separated into progenitor cells and injected into the myocardium. These cells have been teased out of the bone marrow, which is thought to be the nursery, propagated, and injected into either the muscle, or into the coronary supplying the damage muscle. It is still early to assess the results but the BOOST trial demonstrated at least some efficacy in improving left ventricular function. This new means of therapy called cellular therapy represents a new addition to standard medicines.

In the early days of medicine, drugs were rather caustic and would either induce vomiting, cause diarrhea, or act as a cathartic. This is where the notion originated that a good medicine was evil tasting. Heavy metals were injected to help cure diseases such as syphilis. The toxicity of the cures was often worse than the disease itself. Drug therapy became popular with the ability to give compounds that would affect the physiology of the organism. Diuretics were able to stimulate urine production by the kidney. Blood pressure medicines would cause vasodilatation of the blood vessels. Antibiotics became available when penicillin mold was noted to suppress bacterial growth. Hormonal therapy was the next exciting development and allowed diabetics of juvenile onset type I to survive by giving insulin that was prepared from the pancreas of other animals.

Mimicking peptides and other hormones of the body has become quite profitable as noted by Amgen, Genetech, and other biotechnical companies. Medications such as tissue plasminogen activator, made by our endothelial cells can be used to dissolve blood clots. Hormones produced by the kidney can increase the growth of red blood cells. These substances are produced naturally by the body. Scientists have been able to identify these small molecules and learn their properties. They have even learned how to make theses molecules in large quantities by placing the proper DNA sequence into an E. Coli. bacteria. The fancy proteins can be grown much like beer is fermented.

The next major frontier in medicine will be cellular replacement therapy. The controversy over this therapy is the source of the cells. Embryonic stem cells are perfect for uncovering the secrets of cellular differentiation, rejuvenation, and repair. Since this involves a potential human life there has been great controversy over using these embryonic cells. Some countries that do not have stringent regulations have made great strides in this area opening doors to treat degenerative diseases like Parkinson's disease, dementia, ALS, Spinal cord injuries and a host of others.

The controversies in the use of stem cells reverberate around political circles and raise many difficult to answer ethical questions. Do not lose faith. Each of us still has our own embryonic stem cells from our original conception. We can avoid the controversies of cloning and sacrificing potential human life to save other individuals by concentrating on our own self derived stem cells.

As a physician for the last 30 years I have made countless observations. My older patients who are active remained healthy. Younger patients who have afflictions such as degenerative arthritis or obesity that prevents them from being active quickly deteriorate. The difference between these two individuals is the ability to rejuvenate their tissues. What stimulates our stem cells to rejuvenate damaged organs? Degenerative arthritis, diabetes, obesity and vascular disease all are state of inflammation. Inflammation decreases our rejuvenating stem cells.

In 1772, Heberden described the benefit of exercise in patients who suffer from angina pectoris. Heberden was an expert since he was the first to name the disease. Some accounts of a disorder of the breast Med Trans Coll Physicians (London). 1772; 2:59-62. Heberden described a patient who was afflicted with angina and cured himself by sawing wood for 30 minutes per day. He did not know of stem cells but was able to deduce the cure. Sawing wood likely increased circulating stem cells.

In an editorial written by Dr. Paul D. Thompson entitled Arterioscler Thromb Vasc Biol. Aug 2003; 1319-1321 summarized the treatment benefits of exercise in the Editorial Exercise and Physical Activity in the Prevention and Treatment of Atherosclerotic Cardiovascular Disease. He was referring to a recently released scientific statement of the American Heart Association Counsels of Clinical Cardiology and Nutrition, Physical Activity and Metabolism. This committee reviews the literature and makes recommendations based on clinical evidence from these trials. This information is considered evidence based medicine and is presented below.

Lipids

Meta-analysis of 52 trials of exercise lasting greater than 12 weeks included 4700 subjects with the following beneficial changes in lipid:
HDL increased 4.6%
Triglycerides decreased 2.7%
LDL decreased 5%

Blood Pressure

Meta-analysis of 44 trials with 2674 subjects had blood pressure fall in both normotensive and hypertensive patients as follows:
Normotensive fell 2.6 mmHg systolic and 1.8 mmHg diastolic
Hypertensive fell 7.4 mmHg systolic and 5.8 diastolic

Diabetes

Meta-analysis of 9 trials in 337 type II diabetics decreased HgbA1c .5 to 1% and allowed the number of medications to control, sugar to be decreased.
In regard to prevention the Diabetes Prevention Program demonstrated that a 4 Kg decrease in weight and a 593-Kcal increase in energy expenditure reduced the progression to type II diabetes by 58%. This was better than taking Metformin a drug for treatment of diabetes that also reduces progression to diabetes.

Obesity

Three thousand patients enrolled in the National Weight Control Registry demonstrated that exercise and energy expenditure of weekly energy of 2445 Kcal in women and 3298 Kcal in men was successful in maintaining a 30 Kg weight loss for more than five years.

Cardiovascular disease

A meta-analysis of 51 trials of 8440 patients demonstrated a 31% reduction of mortality in patients undergoing cardiac rehabilitation.

How does exercise compare to the high technology of placing balloons and stents into diseased blood vessels? This question was answered by Rainer Hambrecht et al in an article published in Circulation. 2004; 109:1371-1378 Percutaneous Coronary Angioplasty Compared With Exercise Training in Patients with Stable Coronary Artery disease. He did a study to determine the risk or benefit of exercise as compared to revascularization. Not only was it more cost effective for exercise over intervention, survival was better 88% for exercise and only 70% for intervention. The study involved the randomization of patients with known coronary disease at cardiac catheterization to exercise for a year versus angioplasty.

Exercise in women was investigated by Joann E. Manson et al in The New England Journal of Medicine. Vol. 341:650-658 A Prospective Study of Walking as Compared with vigorous Exercise in the Prevention of coronary Heart Disease in women. In women the more energy expended, the greater the exercise intensity, and the lower the coronary events. Other studies have shown that cancer deaths also fall with more activity. The new guidelines for women's cardiovascular health recommends for weight loss 60 to 90 minutes of exercise be performed on preferably all days of the week

The elderly population was investigated in the HALE project and SENECA and FINE studies that drew importance to the Mediterranean diet and exercise. Exercise in the elderly population decreased deaths from cancer, coronary heart disease, cardiovascular disease and all causes. There is no population that does not get tremendous benefit and protection by daily exercise.

The amount of exercise is controversial. The recommendation for exercise has increased over time. Initially, the recommendations were to exercise 30 minutes three times per week. Studies have demonstrated that regression of coronary atherosclerosis takes greater energy expenditure. The minimal recommendation is to perform more than one hour of exercise per day. President Eisenhower's doctor, Dr. Paul Dudley White, a famous Boston cardiologist recommended that one hour of exercise should be performed every day. Dr. White was an avid cyclist and a promoter of exercise and was an advocate for exercise. Exercise should be in everyone's lifestyle including the young, elderly, men and women, and especially those who have inability to do exercise. For those who can not exercise I would recommend a rocking chair and rocking for at least one hour a day.
.

Our stem cells reside in multiple places within our body. A likely residence is in our cellular active bone marrow. They also exist in individual organ systems as niches ready to replaced aged cells. The bone marrow sends progenitor cells to the endothelium (the inside lining of the blood vessel) where they wait to enter the circulation. The mechanism of entry into the circulation from the endothelial storage is unknown. This release may be cytokine related (signaling) or by simple mechanical movement into the circulation by exercise, shear forces, or random release. Once released, the cells will circulate until they reach tissue or endothelium the target of repair. If one accepts the concept of mechanical release of these cells a number of mechanisms come to mind that may improve rejuvenation. The first of these is simple exercise. Exercise will increase blood flow and thus increase the shear force along the endothelial border. The mechanical shaking of blood vessels vibrating at resonate frequencies will shake loose stem cells attached to the endothelial layer. Simple mechanical movement such as walking has an important role in circulating rejuvenating stem cells.

Vascular disease patients have endothelial dysfunction and fewer circulating progenitor cells. Symptoms of angina, claudication, or congestive heart failure may limit their exercise capacity. Simple inactivity may cause a reduction in circulating progenitor cells by less mechanical stimulation. Supporting evidence includes exercise increasing circulating progenitor cells, improving endothelial function. The older individual has fewer circulating progenitor cells and more vascular disease than younger individual. Decreased numbers of circulating endothelial progenitor cells is associated with greater cardiovascular death, cardiovascular events and hospitalizations.

If we study the circulating stem cells and the effects of exercise, we find that exercise increases the number of circulating stem cells supporting this theory. We have investigated EECP enhanced extracorpeal counter pulsation. EECP (Enhanced Extracorpeal Counter Pulsation) is a device therapy for the treatment of angina and congestive heart Failure. The mechanism of this benefit is not clearly understood. EECP consists of timed, sequential cuff inflation from the calf, lower thigh, upper thigh, and buttocks. The inflation is triggered to diastole. There is hemodynamic benefit similar to an intra aortic balloon pump increasing coronary perfusion pressure and decreasing afterload. It is different from a balloon pump in that venous return is also enhanced. The treatment consists of 35 one-hour therapies. The benefit of EECP extends beyond the time period of this acute hemodynamic benefit. The sustained relief of angina and functional capacity improvement has been reported. The mechanism of this sustained benefit is not understood.

The hypothesis is EECP increases circulating progenitor cells. Our conjecture is that the bone marrow sends progenitor cells to the endothelium where they wait to enter the circulation. The mechanism of entry into the circulation from the endothelial storage is unknown. This release may be cytokine related (signaling) or by simple mechanical movement into the circulation by exercise, shear forces, or random release. Once released, the cells circulate until they reach tissue or endothelium that is in need of repair. EECP causes a reversal of flow in the great vessels and this induces strong shear forces at the endothelial boundary layer. The shear force mechanically forces the cells from the endothelial surface into the circulation making them available for replacement of old endothelial cells, repairing damaged tissue and promoting new blood vessels.

The advantage of exercise is clearly not just due to the conditioning. The benefit may simply be the motion and vibration that accompanies walking, running, and other forms of physical exercise. It has been very difficult to measure the profits of exercise in physiological terms. The physiologic terms include oxygen consumption, heart rate, and blood pressure. The help is clear when we look at other endpoints such as myocardial infarction, stroke, cancer, hospitalizations, and longevity. There may be other means of increasing stem cells such as sitting in a vibrating chair tuned to the resonate frequency of the vasculature. Massage therapy, and acupuncture could promote rejuvenation by increasing the circulation of stem cells.

Why is Lance Armstrong the best Tour de France road cyclist in the world? If you compared his heart and physiological parameters there would be little difference among the world class cyclists. A measurable difference can not be found. They will all demonstrate enlarged hearts, resting bradycardia and efficient oxygen consumption. What makes Lance stand out is the same reason he had cancer. I suspect he has high numbers of circulating stem cells that are able to rejuvenate him faster than his counterparts. He can finish his race and repair his sore muscles and tendons more efficiently. He will race better the next day then the un-repaired world- class athletes. Unfortunately, having a high number of circulating stem cells gives a greater chance of one going astray and becoming a progenitor for a testicular cancer.

Ponce De Leon was looking for the fountain of youth. He did not realize he had found the answer. It was not the fountain, but all the walking he did in Florida looking for rejuvenation. Our patients need to learn the lesson of Ponce De Leon. They need to exercise one hour per day to get their stem cells circulating and rejuvenating. In the next 20 encounters, evaluate those patients in terms of physical activity and make an assessment as to their health. The differentiation is clear. Anyone with a walker, obese with degenerative knees, complaints of angina or heart failure that limits activity are individuals in declining health. The most important point taken from the book Take Heart is Exercise, Exercise, and Exercise.

Justice was 88 year-old male with bronchiectasis and intermittent atrial fibrillation. Justice came to the cardiology clinic clear of mind, but poor in spirit. He wasn't performing any activity and when he did exercise he became uncomfortably breathless and began to cough. During the course of his interview with his physician he was informed that his physician made a New Year's resolution for him to perform one hour of exercise a day. The physician suggested that he obtain a stationary bicycle and begin cycling at short intervals eventually increasing to a total of one hour a day. He was told he could break up his exercise into small portions. The next month he returned for his routine visit and told his physician that he could only go five minutes on a bicycle until he had to quit. The physician suggested that he decrease the resistance of the bicycle and everyday attempt to go one minute further. One year later at age 89 he presented back his physician in a fit and especially vigorous manner. He was successful in increasing his exercise gradual manner, bicycling up to 30 minutes two times daily. He noted that when he would get off the bicycle he would be somewhat wobbly in his gait so he decided to take an additional 30 minute walk. He then also began using lightweights to increase his upper body strength. Over the course of the year he had lost 15 pounds and was now as fit as a drum. His outlook on life had changed tremendously. He was happy. He looked

70, and acted even younger. The physician told them he would like to take his picture and turn him into a poster child for the benefits of exercise in the elderly.

Cecile was a 93 year-old farm girl who lived by her self for the last 25 years. She had family nearby who could check on her. Their visits were usually social because her health had previously been quite good. The family now was taking turns visiting every morning to be sure she had made it through the night without falling or getting stuck in the bathroom. Because of her decreased mobility over the last two years she slowly gained weight. The weight gain and degeneration of her spine caused her to be unstable in her gait. She had been trained to use a walker but really didn't wish to use a walker in public because of its appearance. She did use a walker in the house because she knew a fall would result in disaster requiring EMS to get her off the floor. She made yearly visits to the cardiologist because of aortic stenosis which had continued to progress a very slow manner. She had not manifested any of the symptoms of pass outs, chest pain, or heart failure that would require the valve to be replaced. The cardiologist told her of his New Year's resolution to have all of his patients exercise for at least one hour a day. She looked at her daughters with an incredulous look and asked the cardiologist how he proposed she should get this exercise. The cardiologist asked if she had a rocking chair. She said that she had an old good one that sat in the corner. The cardiologist suggested that instead of sitting in a rocking chair that she should actually rock for one hour a day using her quadriceps and calf muscles. She could spend that time reading the paper, listening to the radio, or doing her needlework. Two months later her wit and twinkle in her eyes had returned and she had lost one half pound. She was faithfully rocking in a chair every day and had told her friends in the senior center of this activity. Rocking chairs haven't been studied as an exercise tool, but maybe it should be studied.

The Houckster contemplated his day. It was early to the hospital for rounds with probably 13 hours of work ahead. He anticipated at least one and one half hours of chemistry and math with his lovely daughter during the evening. If he was going to prescribe exercise he needed to do it himself. He started getting up at five o'clock in the morning to do 40 to 45 minutes of bicycle exercise and on alternate days incorporating light weights into his regime. He began doing an additional 30 minutes of stationary bicycle in evening to get to at least one hour. He needed to do this amount of exercise or more if he expected his patients to do the same. His lovely wife then enrolled him in a Masters swimming program in the hope of collecting insurance money. The Houckster was quite certain that his young 24 year-old female swimming coach was being paid by his wife to push him to the limits of his life insurance.

The pool was outside. The pool was heated but it still took a great deal of courage to climb in and out of the pool at 40° F. The initial swimming experience was very good for the freestyle, but when the backstroke or butterfly was taught; the Houckster felt that he was drowning. His boxer shorts seemed to pull him to the bottom of the pool. Part of this problem was cured by buying jammers. His other team members were half is age and offered support and encouragement. It helped that most of them were female with a pleasant female swimming appearance. Eventually after seven months the strokes became easier and drowning was less often. During the course of those seven months the

Houckster lost 15 pounds and turned an additional 10 pounds of fat into muscle. He was sleeping better, joint pains disappeared and he was able to eat whatever he desired without gaining weight. Energy expenditure was increased by nearly 2000 calories a day. Swimming is a wonderful sport. Swimming is necessary if you want to eat the way most Americans eat. Exercise, Exercise, and Exercise.

Mary Jo monitored Murray's heart rate while he walked on the treadmill, used an arm crank, and lifted light weights during his six weeks of therapy. Murray went from being breathless with very little activity to his current state of competitiveness. He was now matching step for step with Joe, a 40 year old recent myocardial infarction patient. Joe had a relatively small infarct and progressed beyond simple walking. Joe and Murray would push each other during their hour-long session. Mary Jo would have to intervene when she thought the heart rates were too high.

Murray turned to Mary Jo as he was leaving and asked. "Why do I have to keep taking these expensive medications? I think I want a drug holiday.

Murray knew he was in Trouble with a capital "T" as he observed Mary Jo's cheeks turn from a pleasant pink to a hot red. The lecture that came for the next fifteen minutes emphasized how lucky Murray was to have responded so nicely to his medical therapy. Mary Jo had seen other patients have a similar response and stop their life sustaining medication and wind up back in the hospital. The heart failure treatments had to start from scratch and would still take 4 to 5 months for the patient to respond favorably. Some did not make it back. The ignorance of patients and their medication was a pet peeve of Mary Jo and she blasted Murray. Murray was smart enough to suck it up.

He did wonder how the drugs got to market and why they were so expensive. Some of what Murray learned is in the next chapter.

Chapter 7
Clinical Trials, Knowledge, and Drug Companies

Advances in medical technology and drug therapies are staggering. Within the last 50 years the death rate from myocardial infarction has fallen from 50% to approximately 5%. Death from congestive heart failure has also been improved. The advances are due to investments in research in atherosclerosis, lipids, and hypertension. Cancer research has also made tremendous advancements. Cardiovascular disease and cancer have many common traits. While some doctors such as oncologists wish to limit the blood vessel growth of tumors, cardiologists aim to promote new blood vessel growth into ischemic myocardium. The research investments have come from government-sponsored trials and industry. While most medical advances are a result of private industry and private funding, the industry sells these advances to the public and thus benefit by regaining their investments. There is no doubt, these drugs have raised enormous profits for the drug companies and that it will take huge sums of money to keep the pipeline going. The orphan diseases get little support since the chance to make profits is limited. The profit in orphan disease is limited because the consumer population is small. The development of therapies for these orphan diseases will still benefit all of us because it advances science and techniques for solving biologic problems.

Drug companies remain a target of criticism for the exorbitant rates they charge patients. Research requires diligent work and clinical approval of new drugs is a very expensive proposition. It is estimated that for every 5,000 new drugs developed, only one will reach the marketplace. The money spent for clinical trials proving safety and efficacy have become increasingly expensive. The number of documents supporting a drug's safety and efficacy could fill several rooms. The FDA requires numerous documented studies to obtain the clinical results necessary for the drug to be deemed safe for the population. A majority of the clinical trials are completed in foreign land where regulations on the trials are less strict. The fewer restrictions on foreign trials allow new therapies to individuals who have no other options. It also exposes them to potential risk of the new therapy.

The cost of a drug is determined by managers, not by researchers or physicians. The managers attempt to recoup losses for many of the failed drugs and to provide money that can be used to develop new drugs in the pipeline. The profitable life of a drug is short since the patent will often expire within 7-14 years of approval. The legal system considerably adds to the cost because every product development has potential recall and legal fees due to possible idiosyncratic reactions. The medical liability is high. Drugs are tested in thousands and even tens of thousands of patients. When the drug is released it is exposed to millions of patients. There is always a possibility that .01% of the population may have a metabolic mutation that makes the drug unsafe for them. This will not be determined in a small trial. The number of potentially affected individuals at .01% is 1 in 10,000. A statistically significant relationship will not be found in the trials and the drug will be approved. However, after the drug is released in 10 million treated patients there will be 100 individuals harmed. If the publicity is plotted correctly and the lawyers are astute, the beneficial drug could be removed from 999,900 individuals who benefited from the drug because of perceived harm and media frenzy. Drug companies are easy targets and therefore attract lawyers.

Medicinal research is continually evolving. Unfortunately, not all remedies are helpful: some are flawed and thus result in malpractice. Before medicine was written down, observations from one physician would be passed down to other physicians with remedies that were thought to be helpful. However, rather than proving to be helpful, some remedies became forms of malpractice. Even Hippocrates, the father of medicine, had flaws. An ancient Greek tribe was known for its loose-jointed physique. In which the skin around the joints appeared similar to burns. The tribe suffered from an inherited collagen disorder known as Ehrlos Danlos Syndrome in which the collagen is elastic and weak. Individuals afflicted with this disorder could become contortionists in a circus. Hippocrates thought the joints were placed back into position by burning. For centuries it was taught that burning amber was to be placed on the skin to relocate the joint. Coincidently, this could be viewed as an effective therapy since as the patient attempted to get away from the pain of the ember, the joint would occasionally fly back into proper position. Obviously, this was not a good practice of medicine.

The key development in medicine was the ability to share written information allowing therapies to be criticized. The opinions of some physicians were highly valued, but science and impartiality was needed to determine the benefits and risks of any procedure. The loudest physician was considered the authority on the subject. Putting arguments onto written paper removed emotion from the facts.

The acceleration of medicinal therapies did not occur until the 1940s with the advent of anti-microbial medications. In 1929 Alexander Fleming first observed inhibition of bacterial growth by a mold. This observation eventually became useful more than 10 years later in 1941 with the isolation of penicillin and proof it could cure bacterial infections in mice without killing the mice. This eventually saved many lives in World War II- including the life of my father. At this time, Informed consent and experimentation was not known to science. Patients were subjected to therapies based on their physician's beliefs. The ethics of this dilemma was simply to trust in your physician.

Physicians carefully recorded signs, symptoms, and laboratory tests determining the natural history of disease. The medical community became aware of the serious consequences of hypertension. There were no symptoms of hypertension early in the patient's life and blood pressure was not recorded until the early 1900s. Following these individuals over time demonstrated how this silent disease could be deadly. Therapy at that time was a diet consisting of salt-free rice and fruit. The avoidance of salt and increase of potassium in the diet could improve blood pressure. Although Franklin Roosevelt was one of the individuals placed on this diet, he still suffered a significant stroke. In the 1960s, simple diuretic therapy was developed and the first clinical trial in hypertension was performed amongst veterans. In addition to caring for war weary veterans, the veteran administration was to advance medical knowledge. One of the first landmark trials in hypertension was performed by the veteran administration hospitals.

The veterans subjected to the first clinical trials attempting to exhibit the control of blood pressure with medication demonstrated favorable outcomes.

The trial was performed in 1967 and was called the VA Trial in Severe Hypertension. There were 143 subjects enrolled in the study who were treated with hydrochlorothiazide, reserpine, or hydralazine. The population treated with the previously mentioned drugs proved to have a therapeutic benefit in the subject group whose blood pressure diastolic was > 115 < 129 mmHg. There were four deaths in the placebo group and no deaths in the treatment groups. This was the first study that demonstrated that blood pressure treatment could save lives. It is notable that the approval of medications was not dependent upon safety but in efficacy of their therapies. Most of the very early medications we use today have never proven to be safe in population trials. However, in later hypertensive trials the number in the trial increased to 19,000 as in the HOT study (Hypertensive Optimal Treatment Study) that demonstrated that a 4mmHg difference in diastolic blood pressure led to a better outcome by 50% in the group treated to 80 versus 90.

The trials performed today involve thousands, and occasionally tens of thousands of patients. The drugs need to demonstrate safety in the population, efficacy in the treatment affects, and also positive outcomes in regard to clinical endpoints such as death, stroke, myocardial infarction, hospitalizations, and general misery. The cost of these trials can be imagined if one considers 30,000 patients each receiving the medication or a placebo. If we assume 8 visits throughout a four-year study, four tests per visit, each test costing $70. The monitoring of these patients would cost 30,000 x 8 x 4 x $70 = $67,200,000. In addition to this cost, one must add the cost of monitoring personnel time, data collection and tracking, any additional costs necessary to investigate abnormal laboratory tests performed to the benefit of the experimental subject, and the rent and upkeep of the monitoring facility. The cost is still raised further by the screening process. To enroll 30,000 patients, 100,000 patients may be screened to see if they meet entry requirements. The hundred thousand patients will require screening tests and evaluations to see if they are appropriate subjects. Expert panels and safety monitors will have expenses as they monitor the trials progress. Exit strategies to inform patients of their benefit or harm from the medications are required with transition back to their primary care physicians. The cost of a single large trial can average $100 million. Industry says the average cost is $359 million dollars (1993) for one drug approval, which is pretty close to my poor estimate. There may be additional need for prior dosing trials and follow-up trials to obtain final approval of the medication. At any time during this process the drug may be determined to fail. The average patient has no understanding of the great expense it takes to approve medications. The pipeline for new drugs shrinks each time a new regulation or concern is raised.

Cost of Drugs
The price of a drug depends on its value to the patient and the ability to recoup expense. There are now 60 million patients taking Statin medication. If we assume a company increased its sales by 2,000,000 each year for 7 years there would be 56 million (2+4+6+8+10+12+14) x 365 doses/7years. It is a good estimate that one quarter of the

doses would be samples or indigent gifts. The number of pills sold would be 107,310,000,000. At a 1-cent profit per pill this would still bring in $1 trillion dollars over 7 years. The math and estimates may be flawed but the profit is certainly large. The competition will reduce the profit because it will be split among other drug companies who have a competing drug. You can test the validity of the numbers by finding the average quarterly profits and multiplying by 28 quarters.

Drug companies are fairly confident before they initiate a trial that the outcome is known. Smaller, less expensive trials will point the way. If the outcome is doubtful, it is unlikely the investment will be made to continue the trial and the drug will be placed back on the shelf. Statisticians can estimate the outcomes of large trials based on information obtained from smaller trials. Unfortunately, large trials in a slightly different population can have an unexpected result. Trials are never risk free. Many times a drug can be studied for one affect and it is marketed for an entirely different reason. Proscar was to protect against prostate cancer as well as reduce bathroom trips and dribbling into the underwear. During the trial it was found to grow hair and was marketed for this side affect.

A successful trial can result in billions of dollars in profit to the drug company. Miscalculating the dose of the medication can result in failure for a drug company. Small molecules that will temporarily poison platelets during non-Q wave myocardial infarction have improved outcomes in patients undergoing angioplasty and acute coronary syndromes. The platelets attach to portions of the blood vessel where the endothelium has been disrupted. The platelets then attract other platelets eventually forming a clump and promoting a thrombosis within the blood vessel. There have been several medications that inhibit platelet aggregation that made it to the marketplace. One of the medications was Integrelin, the other medication was Aggrastat. The dosing of Integrelin was more effective in acute angioplasty, and as a result took over the marketplace. The drugs appear to have similar efficacy with the only difference being the dosage strategy. The gamble that paid off for the manufacturers of Integrelin was having a bigger dose up front in the treatment of acute coronary syndromes without raising bleeding tendencies. Aggrastat took a more conservative dosage approach and lost the bet. Integrelin's bet on getting more of the drug upfront earlier in the disease course was logical and resulted in an economical bonanza. The miscalculation of an effective drug dosage cost the makers of Aggrastat billions of dollars.

Placebo Tacebo
The horse-drawn wagon came up the dusty road and was preceded only by the sound of the jingling bells attached to the wagon. The bells signaled to customers the medicine man had arrived. Inside his wagon would be an assortment of potions, which could cure a number of maladies from arthritis to poor appetite. Entertaining with an assortment of magic tricks and testimonials from previously paid customers would draw the interest of the crowds. The medicine man would be successful in selling his wares if he appeared genuine. Twenty five percent of patients who took these medicines did improve. This power of healing based solely on the power of suggestion is known as the placebo effect. The tacebo effect is opposite to placebo making you worse from your present condition.

When a bad outcome is suggested, 25% of people taking these medications will get worse. The medicine man could use this affect by suggesting, as he was packing up to leave, that if you did not buy his product your luck would likely get worse.

When medications were developed it became quite obvious there was difficulty in demonstrating the medication was helpful. Consequently, clinical trials were born into medicine. The purpose of these trials was to statistically demonstrate the medication performed better than compared to a placebo. It was expected that some individuals taking placebo would get better as did patients taking the true medication. Statistics would verify that the medication was better than placebo if more individuals in that group improved as compared to the group receiving the fake medication.

Global Utilization of Streptokinase and tPA for Occluded Arteries, or GUSTO I, was a trial that involved 41,021 patients in a trial that compared the cheap clot busting agent streptokinase to genetically engineered naturally occurring clot buster that cost greater than $2,000 per dose. The results demonstrated that tPA and heparin administration was superior with respect to mortality 9.1% as compared to the streptokinase group which was 10.1% . The price paid was more hemorrhagic strokes .3% versus .2%. This meant 1 patient out of 100 treated patients with TPA instead of streptokinase would benefit. This was a statistically significant number, and the practice in this country immediately switched from the $200 drug to the $2,000 drug. In Europe, streptokinase continued as a front-line medication for the treatment of acute ST segment elevation myocardial infarction. The gamble paid off for Genetech and it became one of the very first biotech companies and continues to be a leader in the field. The one in a hundred benefit seems marginal, however, if we apply this to the 2 million myocardial infarction's per year the number of lives saved due to this new and expensive drug could be 20,000 lives. The thrombolytic wars continued through many trials. As the difference in treatment affects lessened, the greater number of patients had to be incorporated into the trials. The winner of these trials was TNK, another product of Genetech. The war appears to be over with no competing drugs in the wings. The expense of the next drug war is prohibited and protects Genetechs gains.

TNK was one of 1,000 different mutations of the tPA molecule. It is interesting that a runner up, Reteplace, was sold to a competing company and for a while market share was lost to this molecule. The medication was marketed as a one size fits all and this strategy caused its demise. It did not make sense for a 90 pound Arnett to receive the same dose as a 280 pound Arnold.

The importance of a trial is reflected by the number of patients within the trial. The size of the trial is a guess made by statisticians based on assumptions of efficacy of the medications. It was not a random choice to pick 41,000 patients to be in the GUSTO I trial. The statisticians estimated that it would take this number of patients to prove statistical significance. The greater the treatment difference, the fewer the number of patients need to be treated in a trial to show a statistical significant outcome. In the first clinical trial of treatment of hypertension in the VA Trial in Severe Hypertension, 146 patients were enrolled to show a treatment benefit. The benefit, however, was only observed in the patients who had the highest levels of hypertension. Patients who had more moderate hypertension did not have clinical events because they were less sick. The trial was stopped short because of the outcome in the highest blood pressure group. The patients in the moderate level hypertension group did not demonstrate a treatment benefit during the time of the trial. The answer in this group remained unknown until many other hypertensive trials were completed. The number of patients enrolled in these later hypertensive trials have also climbed. It is much more difficult to demonstrate a treatment benefit in a healthy population than in a sick population simply because the sick population has more events and fewer patients needed to be treated to demonstrate statistical significance.

The number of patients needed to prove a result is certainly important but may not be as important as asking the proper question. To the chagrin of many patients and vitamin advocates, the antioxidants such as vitamin E have not demonstrated a significant effect in the clinical trials. There is fairly good basic science behind the potential benefit of anti-oxidant medications. The free radicals formed within our bodies oxidizing cholesterol particles can damage macrophages and are the initial step in production of an atherosclerotic bubble.

To answer the question, the Heart Outcomes Prevention Evaluation (HOPE) trial was conducted. This was an industry sponsored trial that was successful in demonstrating that the Ace inhibitor Ramipril, a blood pressure medication, was successful in reducing cardiovascular deaths 25%, stroke by 32%, and nonfatal MI by 20%. There were more than 9,000 individuals studied who were at risk for cardiovascular disease but did not have previous events. More importantly these individuals did not even have hypertension by standards of the day. The reduction in blood pressure with Ramipril was 3 to 4 mm of mercury. At the time of this study the slight drop in blood pressure did not seem so significant. Later studies confirmed this small drop in blood pressure by any means has better outcome. This industry trial was unique in that it also tested the antioxidant effect of vitamin E. Over the 4.5 years of study no benefit from vitamin E could be discerned. The endpoints chosen were acute cardiovascular events such as myocardial infarction, stroke. Industry should be applauded for attempting to answer this question. It is one that has no economic advantage to the drug company and was done for science. The government would not commit taxpayer dollars to a study of vitamin E simply because of budget consideration. The investigators who wanted to answer this question did an excellent job of convincing industry to fill in the information void with an experiment that would not add profits to the company. Industry should be applauded.

There is no reason to believe that vitamin E has anything to do with prevention of plaque rupture, a decrease in inflammation, or an effect on the platelets that are associated with acute myocardial infarct and stroke. The science behind vitamin E is that it is beneficial during the early phases of atherosclerosis preventing the cholesterol buildup. There is nothing to suggest that vitamin E would reduce any cardiovascular event studied over a short period of time. The trial that needs to be performed to answer this question is a trial that examines the progression of atherosclerosis over many years. The proper question to answer is: Does early use of vitamin E diminish the intimal thickness of the carotid over time as compared to not taking vitamin E? The study would need to enroll 20 year-old patients and follow them for 20 years. Continued follow up of this group compared to the control over the next 40 years when they begin to have events should give a definitive answer. For now we can only say it will not prevent a myocardial infarction in the next 4.5 years.

Folic acid has a similar story as Vitamin E. Elevated Homocysteine levels have been associated with a build up of atherosclerosis and the level of Homocysteine in the blood drops with administration of folic acid. There is no reason to think that a short-term trial of folic acid would reduce events. This is another example of a large clinical trial that asks the wrong question. The folic acid additive demands a proper trial looking at progression of atherosclerosis is necessary, not a trial of acute vessel closure. More confusion was added to folic acid when it was found that restenosis of stents may be accelerated by this simple vitamin through healing and proliferation of cellular matrix. Therapies when taken out of context can add to confusion especially when multiple mechanisms of risk and benefit are present.

Is it time to replace clinical trials? Currently these trials are the only avenue to approve new therapies. If it takes 40,000 patients to prove a point, is the point worth proving? A more personalized and engineering approach is desired to answering future questions. Cause and effect requires the cause to precede the event, there must be a relationship between the cause and effect with a reasonable explanation for the relationship. All other explanations need to be eliminated. Thinking before a trial is started, modeling the expected result, and doing the trial to see if the model is correct may add to the science. This technique is difficult since the trial will be negative if you did not develop the correct model. It is important to get from point A to point B, but more importantly to know how you got there.

In a further discussion of clinical trials, the correlation between earlobe creases and coronary disease will be considered. How can this relationship be proven? Without a model or any idea of the relationship, a clinical trial can be performed. 100 patients with earlobe creases and 100 patients without creases are selected. The groups will be matched for baseline characteristics with similar cholesterols, blood pressure, smoking history, and other markers so the only true difference between the groups is the presence or absence of an earlobe crease. The patients are then followed for 2 years to see which group has more heart attacks. If 12 in the crease group had heart attacks and 8 in the no crease had heart attacks the hypothesis could be true. The statistician could run the numbers and tell you the probability that this is true. If there was no difference, the trial may not have been conducted long enough since a crease may be a predictor of a heart attack 10 years in the future and the trial was only conducted for 2 years. This trial had inherent bias since the earlobe creases were selected.

A better method would be to take the entire population of Framingham, Massachusetts and examine everyone for ear lobe creases and compare them to those who do not have creases. If we look for differences in the groups we may find there is a difference in something. If we look at 1 parameter and find a difference it may be true. If we look at

100 parameters and find one that is a different it may have occurred due to statistical chance. The simple question is not easily answered if we do not know the relationship between creases and heart attacks.

A model of this problem could be constructed in a manner similar to that discussed here. Ear lobe creases are a result of inflammation. Heart attacks are increased in individuals with inflammation. Earlobe creases may be a marker for an individual who has genetic tendency toward inflammation and is at greater risk for coronary event. In the shopping mall 100 individuals are selected and pictures of their ears are recorded and a free blood test of Hs-CRP is drawn. The Hs-CRP values in ear lobe creases are compared to those without creases (2.8 versus 2.0). Groups from the cardiology, orthopedics, endocrine, internal medicine clinics can also be selected. Some special population will have elevated Hs-CRP and within these populations a difference can be analyzed for with and with out ear lobe creases. Another group composed of consecutive patients arriving in the CCU with chest pain can be selected. If in all of these groups Hs-CRP was higher in creased ear lobes there is likely a relationship between Hs-CRP and creases. An explanation is that inflammation affects collagen and remodels the ear lobe similar to the changes in collagen with degenerative knee arthritis or in the bubble of cholesterol in the vessel. The statistician would determine from the numbers if the relationship were significant. Other explanations should be eliminated. This suggest a plausible model but would need to be taken one additional step and follow the creases with elevated inflammatory markers to see if that group has a greater than expect heart attack rate over the next 10 years. A gene could be sought in those individuals with creases and perhaps a mechanism for the genetic mechanism be determined..

Eventually, the genetic code may be the variable that is tested against a planned drug therapy. This will ensure personalized medications. If you have a certain genetic trait the medication will have a greater probability to work for you as it did in a test subject with the same genetic make up. Sub group analysis will then be possible because of similar genetics.

Generics

Generics are medications that chemically match a trade name medication. After a patent has run out, anyone can produce the drug and sell it for any price they can receive. The only requirement is to demonstrate the drug has the same chemical composition and biologic availability. Most generics will work as well as the parent drug, but the FDA does not require them to be tested against the trade name drug. The trade name drug could be produced much cheaper by the parent company since all of the equipment to produce the drug is already in operation. The trade name company will wait for the pricing of the generic so they can maximize their profit. Restraint of trade laws prevent competitors from dropping the price just to keep competitors from even trying. Until numerous companies compete, the price of generics does not fall a great deal. If a flaw is found in the generic, it will be found by the company that held the original patent. The name brand company will benefit by eliminating the competition.

Research is required of every resident. It should be fun and an idea that is self-generated. I have included some ideas for research that can be fun and also answer some of the questions that I have. The most frustrating aspect of research is the process of having your paper reviewed for publication acceptance. Reviewers have a responsibility to read new research papers and determine if methodology is correct and if the conclusions drawn by the author are correct. In general, a reviewer is deemed to be an expert in a field of research. They are supposed to be unbiased and only look at methodology and science to approve a paper for publication. Reviewers are not unbiased. If they think your idea is nuts they don't waste their valuable time reading any further and just say rejected. A conscientious reviewer will send back the reason for rejection and make suggestions. Most don't bother.

The feeling of euphoria, insight to previously unsolvable problems, is almost as good as sex. When you are rejected by other researchers, it is like being told by your first love to get lost. The ups and downs of research are like a rollercoaster ride. I still remember my first insight. As a first year fellow you were responsible for discussing a topic in cardiology and were expected to obtain original work from microfiche. There was no Google to get 1 million hits on a subject. My topic was aortic stenosis and its hemodynamics. Gorlin and Gorlin had published a paper in 1951 which presented the first cardiology rule in determining the mitral valve area and aortic valve area. It was a formula that all fellows had to understand to pass their boards and more importantly to take care of patients. With my engineering background I found a problem with the formula. The original formula included the term for gravity. The reason for the mistake was a common practice among civil engineers to equate pressure with pressure head the height of a fluid column. The formula as written with the "g" a gravity term would imply that travel into space was impossible since in zero gravity there would be no flow across the mitral valve. I corrected this mistake and sent it to Dr. Gorlin who sent the problem to a NASA chief administrator. Eventually, a corrected formula was published. I was recognized by the statement, "It has come to our attention..." Dr Gorlin wrote me a nice letter. I certainly did not need a mention in the paper, but it would have been nice to demonstrate to my fellows that you must always question your teachers.

Research Ideas

Nutrition
    Investigate the inflammatory effects of food types (I want to write a cook book that lowers inflammation and cures arthritis and heart disease).

Physical Exam
    1. Ear lobe Crease and HS CRP
    2. Systolic pulsation of pulmonary artery and PA pressures greater than 50
    3. Color of palpaebrae and prediction of blood transfusion in post operative CABG
        a. Prediction of blood transfusion post CABG by color chart comparisons to palpaebrae

b. Early transfusion based on palpaebrae reduces post CABG length of stay volume excess, atrial arrhythmias, and mortality

Myocardial Infarction
1. Regenerate infarcted myocardium with GCSF and EECP
   a. EECP is safe 48 hours after cardiac catheterization
   b. Retrospective analysis of EF by echo post MI demonstrates no recovery in EF
2. Develop a tool to measure cardiac regeneration by serial MRI
3. Plasmapheresis to protect injured myocardium and set the stage for cardiac regeneration with EECP
4. Cause inflammation away from the heart as a decoy

Myocardial remodeling
1. MRI assessment of regional hypertrophy comparing LBBB to RBBB. The conduction system determines cardiac remodeling
2. Intermittent RV/ Bi-V pacing   dogs, patients with RV failure, Pulmonary HTN
3. Electrical remodeling of the heart proof of concept
4. Coreg increases EPC's and stem cells greater than metoprolol

Hypertension
1. Fat females with edema – do diuretics cause hypertension?
2. Urinary sodium to potassium ratio selects patients who are salt abusers
3. Dietary instructions in salt abusers can lower blood pressure with the aid of weekly urine tests that ensure compliance
4. Spironolactone used as an initial BP medication in patients presenting with BP>170 will be more effective than traditional titration of medications
5. Measure BP as patients are getting out of the car as compared to after their walk into the office, compared to technician and physician

Congestive Heart Failure
1. Titration package for CHF is more effective than nurse titration
5. Develop a simple color urine test that determines if Urinary sodium is greater than urinary potassium
6. Demonstrate that as needed diuretic use triggered by an abnormal urine sodium to potassium ratio has better renal function and symptoms
7. Retrospective analysis of Spironolactone use in the CHF clinic. Does Spironolactone use have better diastolic function over 4 months as compared to no use of Spironolactone?
8. Retrospective use of Erythropoietin in the CHF Clinic. Are there outcomes improved as compared to matched controls that did not receive Erythropoietin
9. EF change after initiating Erythropoietin prospective trial
10. Hormonal changes in the natural history in patients with CHF- Does multiple endocrine failures explain a patient's decline? What is the normal hormonal status in CHF, Does it change over time,

11. Diaphragmatic training and CHF using a CO_2 charged incentive spirometer
12. CPAP in Decompensated and obese patients
12. Do CRP and BNP correlate in acute decompensated CHF?

EECP

1. EECP and diabetic peripheral neuropathy
2. EECP and stroke
3. EECP and neuromuscular diseases, Parkinson's, ALS
4. EECP and inflammation

Emrx SEARCH

1. Patients presenting with MI on a STATIN which statin are they on and how does this compare to statin use in the population
2. BP control when Norvasc was changed to Plendil
3. How many males present with MI on alpha-blocker compared to the population?
4. BP systolic, diastolic and mean placed into quintiles and determine the potassium level in each quintile i.e. low potassium predicts hypertension
5. Same thing with other electrolytes and creatinine

Population Screening

1. Fill the technology gap for patient's Computer based screening, education, tracking program for wellness
2. Screening for Cardiovascular disease in the population using simple parameters and measurements Abdominal girth, brachial ankle index, BP, three questions for angina, claudication, stroke, lipid and CRP, earlobe crease family history
3. Feedback health risks to the population to affect change in bad health habits closing the Loop of medical information.

Coumadin   Can it be stopped after treatment period for atrial fibrillation

What happens to patients when Coumadin is discontinued? Many patients are treated with Coumadin for atrial fibrillation. Some of them will develop a contraindication for Coumadin due to falls, bleeding or some other reason such as demanding to come off of Coumadin. I want to know what happens to these patients. My hypothesis is that treatment with Coumadin for a period of time will provide continued benefit. I would expect the stroke risk after discontinuing the anti-coagulant would not rise to the historical control. If patients were entered into this registry and data updated by the responsible individual who entered them, my hypothesis could be proved and then a clinical trial could be performed to confirm the finding. Ethically, no one should be taken off of Coumadin. The registry could prove that only temporary treatment is necessary.

Testing

1. Retrospective analysis of testing in patients greater than 250# over the last 3 years. Sensitivity specificity outcomes for testing in the obese

2. Squat, shoe tying as a stress test.
3. Squat to determine volume status of chronic CHF patients. The shorter the time to shortness of breath the higher the filling pressures.

New Model of Disease

Regeneration = Degeneration is health

Regeneration > Degeneration is growth, cancer, restenosis, LVH, Diabetes II etc.

Regeneration < Degeneration is age, arthritis, dementia, Diabetes I etc.

Inflammation is the fulcrum between regeneration and degeneration

Failure to regenerate is secondary to lack of stem cells, lack of a channel, lack of an electrical field, failure of cells to communicate with each other. The default setting of stem cells is to become a fibroblast.

Inflammation is good and bad and very difficult to understand.

These are only a few of my questions that I would like answers. As a physician we all have a responsibility to make observations, analyze trials and arrive at new answers to help our patients. Without research Murray's remarkable recovery would have never have proceeded. A medical system must advance the art of medicine and our system of care will be briefly examined.

Chapter 8
Our Medical System of Delivery is not a Democracy - The Future of Health

Murray had been sitting in the Houckster's Office for ten minutes staring at several bills with a confused, irritated gaze. The Houckster had seen this look before. "Go to the Business Office, Murray." the Houckster quipped. Dr. Houck knew he could not unravel the confusion of multiple insurance companies trying to pay the least for Murray's health. He had TRI Care, a supplemental from the Blues and was on a senior program. The status of his account made Murray think the bill collectors would be out in force.

The Houckster had seen this before and long ago decided he could practice medicine or be a forensic accountant. He elected to ignore accounting and let his wife pay all of the bills. The grey haired physician had a suspicion that many of his seniors actually paid these bills when they were really notices that a rival company had not released the funds. He wondered how much of that money was ever returned to the poor patient who could not stand being accused of having a debt. The Houckster was unhappy with the medical payment system.

The Houckster was fortunate because the physician group had their own health plan and he never had to ask permission to perform any test. Other plans required pre-approval and still would deny payments. How did this mess of a payment system come to be?

The world ahead can be a dangerous place. It is estimated that by the year 2011, 17% of the gross domestic product will be utilized by health care expenditures. The expenditure per individual is estimated to be $9,216, and it is estimated that the current level of care cannot be maintained with a stable economy. (Prof. Thomas E. Getsin. Modeling Long Term Healthcare Cost Trends, The Society of Actuaries December10,2007) It appears that our system of health care is broken. The baby boomers are becoming aged and afflicted with multiple disease processes. This could be our undoing. The baby boomers will demand new artificial knees for their arthritis, mechanical hearts to replace their failing hearts, robotic home health providers to help them shower and perform chores of daily living. They will not be happy when the money runs out and the services end.

If we evaluate our current system of care by examining our care of the sickest patients we would receive straight A's. The health of our populations, however, would not receive favorable marks. In the past our country was the leader in the healthcare industry. The new development of drugs and devices, lifesaving procedures originated within our own country. The world has caught up and many of the innovations are now taking place in foreign countries and even previous Third World countries. The strict regulations of the food and drug administration have forced some of the developments into foreign lands. This has given our people some protection to unscrupulous experimentation. It is also limits the ability of new innovation to help our ill patients.

Our system of health delivery has evolved from an individual caring physician doing the best he can for his community to mighty board run corporations with a profit motive. The old country doctor model, paid with chickens and vegetables for services rendered was not the best economical model for physician family security. Payment is critical to the physician to survive and it was quite unpleasant to collect debts. Many times the debts were simply forgiven. It was much easier for the physician to practice medicine if there was an intermediary between the patient and the physician who would take care of the expenses. The security derived from the insurance system both private and government was a welcomed addition. Physicians could just worry about taking care of patients and not collecting bad debts. Charity work was always a part of medicine. Charity is simply easier if most of your patients could provide payments even if the payments were reduced. The price that physicians paid for this convenience and a small measure of security is freedom to treat his patients in the manner he feels is best.

The insurance systems and method of delivery of health care is not a democracy. Physicians, patients, hospitals, clinics, medical device manufacturers, and drug companies are all in servitude to the small kingdoms provided by the insurance system. Individual physicians have very little say in the management of their patients. The drugs that are selected are preferred drugs based mainly on cost and favorable data. Required guidelines and best practice of care has been a positive evolution. Practice guidelines have removed more autonomy from the physician. The real loser in this current system is the patient.

The current system profits from people becoming ill. I would be cruel to suggest that physicians and insurers wait for illness- it is true. There is no good incentive for patients to stay healthy. People do not come to the doctor when they feel all right. The system completely fails people who did not come to the doctor. With the advent of co-pay's many sick patients do not come to the physicians when they should. Instead, they wait too long and come in sick and cost the entire system a great deal of money. Physicians particularly benefit from people's illness since they are paid by procedures and RVU's. Physicians should be paid to keep people healthy to reduce the cost to the entire system.

Society, including employers, is not taking responsibility for their employee's health. One of the greatest detractors from the bottom-line business profit is the expenditure required for health care. Employers attempt to shift this to patients. The end result is an increase in the number of under insured.

Our wonderful health care system is not a democracy. It is like a 16th century fiefdom with lords, servants, knights who all work for the "man." We get a little bit of security for a big tax load. If we fight the system we get eliminated.

Recounting what we learned in chapter 4: nutrition, inactivity, decrease in calorie expenditure and the increase in refined calories intake has resulted in obesity and new health concerns. More than 50% of the world, a percentage outnumbering the starving population, is obese. Without individual responsibility to curb this epidemic, new demand will be placed on the health-care systems. In our own country and in countries throughout the world, the managers of health care system are not prepared for the influx of patients that will occur due to this epidemic. Without the incentive for the patients to improve their own health, our entire economic system may be driven into the ground due to over-utilization, loss of productive activity, and diversion of resources from other productive areas to maintaining our population's health.

We have tremendous technology to help diagnose and treat sick patients. There is a substantial technology gap. The greatest technology gap is at the level of the patient. Patients have no idea of their risk of cardiovascular disease, cancer, or inherited disorders. It is rare to find individuals who actually have had their blood pressure tested and remember the results. It is even rarer to find individuals who have had blood tests that are crucial in determining their cardiovascular risk: cholesterol profile, hemoglobin A1c, glucose, high sensitivity C-reactive protein. Only a few elite actually utilize this information to help improve their health. Common cheap measurements that can be obtained in the privacy of the home are not performed. The abdominal girth (not waist measurement) of greater than 35 inches in females and greater than 40 in males is a significant risk factor for metabolic syndrome. The abdominal girth is not the size of the belt, but is the measure of the first part of the body that passes through the door.

Patients are not tracked for wellness. Patients have no incentive to exercise and remain healthy. It is difficult to regulate and measure patient exercise. Employers have no incentive to keep their employees healthy, although, it has been demonstrated that productivity can be improved in a happy an energetic healthy working population. Simple advertisements on television to stop smoking and stay active, is not enough for our population to maintain their health. These warnings that come on television have the same effect as your mother telling you not to run with scissors and to never use or shoot a BB gun. The schools have been ineffective in their health programs failing to teach what is necessary for a healthy lifestyle. Schools and employers do not set examples or goals for methods to increase physical activity and to stay healthy.

It is time for *a revolution, a revolution* of health care providers, patients, and employers to overthrow the insurance companies. It is time for the individual patient to take responsibility for their long-term health needs. It is time for the employer to take responsibility for their employees' health. The revolution begins now with a simple premise. I want to be healthy, if I am sick I want the best medicine can offer. I am responsible for family, friends, neighbors, my employees' health, and myself.

A revolution can be started with a thought, a book, a defiant refusal to give their seat away in a crowded bus, or a web page. Violence is often associated with abrupt change. The book that you are now reading is part of a revolution. The web page Houckstertakeheart.com is the tool. On this web page is the initial version of a wellness tracker and personalized medical record. It will suggest best practices and how to keep healthy. It will provide the technology gap at the level of the patient to provide wellness.

The change in health care will be an initial slow process with a rapid revolution. It will have to start with volunteers and no monetary resources. In time, money and resources can potentially flow into this system making it self-sustaining. Eventually it will rob the insurance companies of their premiums because it is a cheaper and better alternative for health maintenance. As more people become part of this program, the wellness program can take on greater responsibilities. Pharmaceuticals can be purchased in bulk and be provided to the members at a reduced cost. Community needs to take care of the community.

People have forgotten about taking care of each other. We have insulated ourselves from individuals that are needy by developing a social system to care for the unfortunate. Taking personal responsibility for our neighbors can be avoided if we let the government do the work with our tax dollars. It seems much easier to provide tax money then provide handouts to help people survive. These handouts are only Band-Aids and do not get to the root of the social problems that we have in our existing communities. The government is too large to react in a timely manner to people's needs. The disaster of Katrina is an excellent example of how the government failed with its huge resources to provide timely care to individual families. Within the caring community of churches and civic groups these individuals were rapidly assisted with housing and food. The thought of who should provide the resources was not entertained, it was just accomplished. Each of us has a responsibility to our neighbors. Social problems require each individual to take a personal interest in improving these problems.

The current malpractice system is certainly inept when individuals can receive $200 million for harm perceived due to a drug that the scientists do not know the mechanism of action. A study to prove this medication harmful would require an experiment involving more than 30,000 individuals. The malpractice system in this country is broken and needs to be replaced. Malpractice in our legal system is not a search for the truth, but a race for the bottom line. Malpractice lawsuits do not improve health care. Physicians who are sued rarely become better physicians because of the lawsuit.

It would be nice if for every bad outcome monetary reward could be given to family members or patients even if malpractice was not considered an issue. The purpose of insurance is to help individuals when unexpected events occur. The determination of damages would come from a blue book value system applied equally across all ethnic and economical backgrounds. Claims would be investigated by an expert panel and not by an unknowledgeable 12 member jury. A panel of experts should investigate a malpractice claim similar to an investigation committee of an aircraft accident. The responsibility of the experts would be to determine the cause and suggest solutions to prevent the event from occurring again. If further training for physicians is required this would be mandated and a physician would be monitored. If an impaired physician were recalcitrant to retraining, they would be eliminated from the practice of medicine. This type of system would improve the quality of care and the quality of physicians. The current system does nothing but adds cost and occasionally gives unreasonable payments to individuals and their lawyers. Malpractice litigation has better odds than winning the lottery.

Lawyers would not be required in the system unless they were trained specifically in quality health care and standards of practice and were selected to a panel of experts. In this way medical care can be improved. The experts would be trained local members of the community. They would also be given oversight to allocate medical resources. Those members would have the responsibility to protect the individual and responsibility to the community to decide where resources will be allocated. The actual change will be difficult to predict since individual communities may wish to approach the great problems of health care in a different manner. There are some communities that will still require outside intervention because they lack resources within their own community. Even these communities however, can be transitioned to self-determination as they become more efficient in their maintenance of health and prevention of disease. I would predict that when the insurance companies sense this change they would become part of the solution rather than dissolving. Insurance companies tend to respond when their patients begin an exit. Without a patient population base they cannot survive.

The Orchestration of a Revolution
Joe, a retired book-keeper and member of a small Catholic parish read the novel *Take Heart*. He doesn't understand all of the ramifications of heart disease and its signs and symptoms but clearly has a revelation that lifestyle, exercise and nutrition is important in health and great value to the community. Healthy people are happier people. He understood how important exercise was to everyone especially the elderly.

In his youth he had been an amateur historian and studied the church's role in the healing arts. Providing medical care and fulfilling religious beliefs often resulted in a conflict between religious teachings and what was good for the patient. One of these conflicts resulted in the separation between medical and surgical specialties. Surgery was taught to barbers by monks when proclamations came from holy powers that medical care was secular and most of their efforts had been drifting from faith to healing the sick. Many of the monasteries became the model for our current hospitals. Faith and medicine was deemed incompatible and barbers performed surgeries because it was felt to be unclean and inappropriate for a priest to perform these surgeries. Opening a skull to relieve pressure from a subdural hematoma could be lifesaving but could also interfere with the passage of the soul to the heavens. Cesarean section to save the mother's life and an infant's life had been debated and essentially solved by instructing barbers to pluck the baby from the abdomen instead of sacrificing the baby's life in the birth canal to save the mother. The barbers were instructed in these a surgical techniques so that that learned priests would be their cognitive counterparts without soiling their hands. This is the basis for our current division between surgery and medicine.

Joe, perhaps inspired by heavenly inspiration, could see the resources of the church being allocated towards improvement of parishioners' health. The church was located generally close to the people it served. The church's space resources with recreational halls were for the most part underutilized except during church school and church social events and suppers. Joe had a treadmill and a stationary bicycle that were collecting clothing in his bathroom. He went and had a discussion with the priest and convinced the priest that he should store his bicycle and treadmill in the recreation hall and make it available to the members of the church. He then advertised a prayer and exercise session. The priest suggested that rocking chairs be provided for those members of the church who were unable to safely get on exercise equipment. He was aware of this because he used a walker because of a prior hip fracture. He knew simply rocking and praying was better than sitting and would likely be of benefit to his parishioners. A donated computer with internet access was placed strategically near the door. As individual church members would come in for exercise and prayer they would be enrolled in the Houckstertakeheart wellness program and virtual primary care physician.

The program asked several fundamental questions and suggested certain blood test that could determine patient's risk for future events. Laboratory blood drawls were provided by local clinic at cost. These values would be entered into the individualized patient database. Additional information could be added including medications, past medical history allergies. The local priest had the opportunity to ask questions about his sermon and suggest changes in behavior. Evaluation of the patient's well-being and compliance to the suggestions could be electronically queried and anonymously made available to the priest and all participants. This feedback is important to evaluating one's effectiveness.

An example follows. The priest had suggested during his homily to greet individuals during the next week with a positive attitude and particularly a big smile. He knew a smile was a tremendous ally. He had learned this lesson prior to becoming married to the church. In high school he was athletic, a bit shy, and extremely short. He lacked confidence because of his short stature. One day he inadvertently smiled at the most beautiful sophomore female. She incredulously, incredibly smiled back. This was observed by other sophomore females who then greeted the short athletic soon-to-be priest with smiles of their own. A smile can be infectious and certainly makes one feel good. A simple smile had returned increased popularity. At the end of the homily the priest had asked his parishioners to greet everyone may next week with a smile and open hand and a helpful attitude.

He then placed a question in the electronic database for his exercise participants to answer. He inquired if they did indeed greet strangers with smiles and an open hand. He then asked several questions about their overall well being and if had been a good week, so-so week, or a bad week. The straw poll results were overwhelming; the individuals who did greet strangers with smiles had an overall positive feeling for the week. Individuals too shy or too forgetful to perform this act did not have as good a week. The parish will be informed of the results and the non-smiling will be encouraged to smile next week. Smile therapy will be part of the priest's homily in the future.

The parishioners during the next week would enter their data into the wellness tracker and virtual primary care physician. The patient would receive a copy of his report. Suggestions for future laboratory tests and possible medications based on their risk would be part of the report generated. Medications, allergies and past medical history would be added to their personalized data. The database could become an individualized medical record and is accessible from anywhere. Recent electrocardiograms could be scanned into the database for ready access in times of emergencies. The easily accessible electronic record could become lifesaving during the time of an emergency. Exercise activity levels would be recorded and tracked. Blood pressures, weight and abdominal girth would be recorded and tracked. Incentive to continue with this activity would be a reduced cost of health insurance. Some of the individuals would be unlucky and the risk assessment would reach a threshold that would require further testing such as an ankle brachial index. If certain symptoms were entered into the database other laboratory tests would be requested such as an exercise treadmill test for the complaint of chest pain, a BNP for the complaint to shortness of breath, a glucose test for thirst and frequent urination. By progressing through screening tests individuals with a high risk could be identified. These individuals would have either pharmaceutical intervention or further testing. Before going to their primary physician the patient would enter their chief complaints and print a copy of their entire medical record complete with problem list medical problems, family history, a review of systems, medications and allergies. In addition the recommended screening tests and immunizations would be displayed for the physician. Cutting and pasting from the Houckstertakeheart web page and entering his physical exam findings and plan the physician will have completed a level four visit. His time can be spent looking for other problems and getting to know his patient better.

Screening all individuals for a single disease entity is not cost-effective. A pyramid approach with testing only those individuals at high risk for the disorder can be a lifesaving and cost-effective method of delivering health care. Simple interventions such as stopping smoking may be all that's required for some individuals to lower their risk for future events. Other individuals, smokers who have symptoms may warrant further testing to diagnose significant coronary blockages that can extend life with surgery.

Slowly, other individuals like Joe will begin wellness programs within their churches, synagogue, or mosque, or community centers. They can use the tools of the web page to enroll patients. Managers of the wellness system will be Medicare aged individuals who volunteer their time. It is only fitting and appropriate that individuals who are high consumers of health care would be the same individuals donating their time to keep everyone healthy. Training programs can be initiated to help patients obtain medications through the available assistance programs.

Training involving chronic disease management programs can accompany the wellness program helping patients understand how and when to take their medications. Calling them to remind them to obtain refills and helping to serve as a resource for simple problems. Small contributions to the local chapters will allow them to expand exercise facilities and begin local insurance funds to help unfortunate individuals. Pressure can be brought upon the insurance companies forcing them to lower rates in individuals who have documented successful improvement in their lifestyle. Documentation would include personalized exercise monitoring measuring activity per week, scales measuring weight changes, breath tests to prove smoking cessation. These patients who are demonstrating healthful changes in lifestyle should be rewarded with lower insurance rates.

Eventually these homegrown wellness programs may wish to fund their own local insurance programs. This bold step would cause the demise of the great insurance companies. Health care would then be more democratic with individual community groups electing leaders to determine what to do with health care dollars. Organization with the help of the web page and remaining decentralized with local money supporting local activities would restore democratic care. Goodbye insurance companies. Goodbye government control. Goodbye needless paperwork. Say hello to a new healthy population.

As this healthful page grows in numbers, the information can be a political force that can force change in our current health care system. The change would be to provide incentive to the individual to stay healthy. The money saved by not becoming sick could then be put back into the local community to improve medical care delivery and wellness. The web page will provide tools and suggestions for the community to develop their own local wellness chapter and optimize their current health care delivery system. Patients with chronic disease can be tracked on this page and again monitored for best practices. The chronic diseases and appropriate medical management of these diseases can be monitored to be sure all individuals were treated according to guidelines. The next stage of the revolution will be the development of community centers where groups within the community will take charge of their health. Individuals, schools, religious centers and government will be on the same page. Every policy that is passed will be scrutinized to see if the policy will help make the community healthier.

Without the above changes medical care will not progress. Medical care is economically driven and is controlled by regulations and billing. The practice of cardiology has pressures to change. The pressures are always economic. We learn from our patients by following them during and after their exacerbation of an event. The event could have been atrial fibrillation, a heart attack, or an angiogram that demonstrated mild disease. Economics of the cardiology clinic does not promote follow-up care. A follow-up level four pays **$ and a new consult level 4 pays **$. The incentive to see patients back is now monetary. Many of the staff and previous fellows have only had three years to follow patients and a great deal of learning still needs to be accomplished because patients tend to teach you things at a slower pace. Ten years or even fifteen years may be necessary to learn the lessons from a single patient. The fellows have continuity clinic that helps them learn follow up care over three years. The things they learn from their patients will be applied for a lifetime. This method of learning may become extinct because it is not a paid endeavor. Without new lessons, advances will come to a halt.

The economics of reimbursement determines the type of care. Incentives to do angiograms and stents will assure more will be performed. Technology is blamed but the real culprit is the dollar. The United States does more surgery and more angioplasty than any other country. Our mortality from cardiovascular death is no different. The procedures are performed in good faith and treatments are within our current guidelines. Patients expect to have angiography and stents. We are paid to accomplish that task.

Linda was a 250 pound legal secretary who was eating lunch and developed a severe pain under her sternum. The pain was excruciating 10/10 and was accompanied by nausea, profuse sweating, and dizziness with shortness of breath and feeling of doom. EMS was summoned and sub-lingual nitroglycerine was given. Vital signs immediately deteriorated to a heart rate of 30 and blood pressure of 70. Fluids, atropine, and oxygen restored vital signs to normal. In the emergency room, the electrocardiogram was normal and cardiac enzymes were normal. Linda was convinced she was dying and the emergency room personnel responded in a professional panic. Despite the good prognosis she was admitted.

Cardiology was consulted and decided to send her for a coronary angiogram. She could not walk a treadmill due to a sore knee and echo images were poor so a Dobutamine non-stress echo was impossible. A thallium could be ordered, but would likely have breast artifact and would require another day in the hospital. She was taken to the cath lab that afternoon. In the mid LAD there was a 70% stenosis with little other disease. The options were to treat medically with Statins, aspirin, beta-blockers, ACEI, exercise and weight loss or to treat with those medications plus Plavix and a medicated stent. Stenting with a medicated stent over a bare metal stent would not improve mortality. It would make restenosis a re-growth of tissue less likely. In fact stenting would not improve mortality. It would reduce angina. Did Linda have angina? Her presentation was dramatic but the symptoms could be explained by esophageal spasm and nitroglycerine syncope. A decision had to be made.

Treating Linda with medications was cheap and would treat the entire vascular system. A stent would treat $1/50,000,000^{th}$ of her system. She still would have to take the same medication to treat the rest of her blood vessels. Linda knows that blockage in the coronary artery is bad. Thinking about the blockage already caused her to hyperventilate with dizziness and tingling around the lips and a feeling of doom. A re-breathing mask was fitted to her face to ward off the terrible feeling of hyperventilation. She wanted a stent.

The cardiologist would be paid $50 a visit. Many visits would be required to get her on the correct medications and to change her lifestyle of eating to excess and not exercising. With each visit the cardiologist would be losing money since the follow-up care is not reimbursed very well. Placing a stent would take less than 15 minutes and would reimburse the cardiologist $500, the hospital for the technical fee $8,000, and the device manufacturer $3,000. Everyone including Linda wanted to give Linda a stent for a total cost of $11,500.

The pressure of using high technology is enormous and the catalyst is the feeling that we are doing the right thing for the patient. Doctors can rationalize the same as anyone else, especially when everyone else is in agreement. The path that has been chosen is a very expensive path. The costs need to be understood. Linda should have a good result with little risk. Risk is, however, present and the risks include the following: perforation of the vessel, dissection of the vessel, early thrombosis, late thrombosis, patient non-compliance with taking Plavix resulting in acute myocardial infarction, and death. Linda may have the feeling that her only blockage is cured and she does not have to change her lifestyle. Linda will have another chest pain if her original complaint was not her heart and was due to esophageal spasm. She will be admitted and an angiogram will be ordered to check the status of her new stent. She may become a cardiac cripple and have many more angiograms each with cost and risk. The simple decision of placing a stent is not simple and financial pressure is on the side of technology and not on the side of doctoring.

Linda had her stent placed. She developed a hematoma that left a painful bruise in her groin and upper thigh but this required no additional therapy. She lost 20 pounds over the next three months on a starvation diet. She did not start exercising. Linda regained her weight and had three subsequent negative angiograms. She decided to start smoking in an attempt to lose weight.

The above case is unfortunately not rare. Coronary stents are wonderful in acute coronary syndromes and acute myocardial infarction. Stents are also a redeemer in patients who already had coronary bypass with failure of their grafts. Most of the time stents are placed in appropriate lesions. Medicated stents have decreased the number of interventions required.

This chapter does not demonstrate medicine at its best. It is a plea to bring changes. Prevention is better than emergency care. Payment for medical care is necessary but also an evil device that removes freedom. The problem with our health care system is the government, insurance companies, the doctors, and the patients. No ONE can fix the problems but WE can!

The Houckster finished Murray's exam and based on a soft S1, a paradoxical S2 and a PMI that was displaced knew that Murray needed a Bi-Ventricular AICD. He gave Murray an order for an echo that would confirm an ejection fraction less than 35% and cardiac dys-synchrony. Murray was also given a consult to visit with an electrophysiologist who would implant the device. Murray's next step to recovery was to live better electronically.

Chapter 9
Predictions - What Makes the Heart Better and Other Nonsense?

Electrical Remodeling

Frankenstein's monster was created by judicious use of a large electrical surge generated by lightening. His life began with electricity. It should not be surprising that the clones that have experienced life began with an electrical shock. Electricity and its' interaction with biology are still poorly understood. There will be controversies over its' use and how it fits into medicine.

During the Revolutionary War, medical controversies were similar to those we are faced today. Prior to the use of antibiotics, groups of physicians believed in bloodletting and other groups of physicians felt that removing blood was a barbaric and senseless act. Bloodletting was performed through simple incisions and heating a glass to form a small vacuum that would suck the blood out of the individual. These devices can be placed on the back or other convenient body parts. Summertime bloodletting utilized leeches to suck the blood from the individual. This technique was recently reinstituted to help save limbs that had been amputated and reattached. The congestion that would occur within the limbs could lead to more tissue damage. These leeches are not the typical ones found in musky riverbeds but are grown in a clean and sterile environment.

Bloodletting could often result in immediate improvement from suffering. Patients suffering from congestive heart failure found relief in bloodletting as it decreased intravascular volume, preload. Unfortunately, once the patient was significantly bled, they would suffer from a new condition-anemia. Bloodletting oftentimes showed immediate improvement but long-term harm.

Medicine is not much different today. We still have physicians raising controversies about established treatments. The proponents may claim to be more scientific by dreaming up clinical trials to prove their points or using meta-analysis to rehash old data. The reality is that when we take care of patients and investigate new scientific ideas we all have our own biases. The practice of medicine is well termed because we all practice and learn from our patients. A good physician will always attempt to find a solution to the patient's problem even if the solution is ahead of the published guidelines.

The following statements are based on predictions, scientific data, and guesswork as most hypothesis originate. The first prediction suggests the electrical conduction system is responsible for remodeling of the heart. The term I will use is electrical remodeling. What does the term electrical remodeling mean? Previous cardiac training involved both electrical conduction system and the mechanical pumping ability to the heart. Combining the two systems in a coordinated manner was never emphasized. The two systems were felt to be equal and separate. This simple demarcation allowed the subspecialty of cardiology electrophysiology to come into existence.

The pumping function of the heart requires the coordination of many cells that line the fibrous skeleton. Over time, the cells will wear out and have to be replaced by stem cells that will differentiate into fibrous skeleton, muscle, blood vessels, and nerves. How the heart regenerates and how the heart develops in an embryo is still a great mystery. This complex organ is built from a bulging tube that twists in a preferred direction and eventually forms the four chambers and valves within the bulging muscle mass. The collagen skeleton may be the roadmap allowing individual cells to be laid down in an organized manner. How does the skeleton know what shape to take? Perhaps electrical impulses attract the cells and teach the stem cells what type of cell it should become. The cells find their home coming out of the circulation to repair and replace aging cells in the heart. A few years ago it was taught and still believed by some that heart cells could not divide and replace themselves when injured. Our own cardiovascular institute director stated, "We are not newts and can not regenerate our tails." The number of cells in the heart did not change after birth and would not be replenished.

Anversa, a wise and freethinking scientist, proposed good evidence that the heart is replenished on a continual basis. While the cells are long lived, they do not accommodate the longevity of the human life, and therefore must be renewed. The replacement cells are small and difficult to find which made them elusive to our scientists. How do these cells find their pathway and maintain the architecture of the heart? One theory holds that the electrical system, consisting of the electrical cellular fibers and electrically active cellular channels, is the unseen architecture that communicates to the cells where they belong. This unseen force causes the differentiation of the stem cells to renew the heart.

The electrical system, I am quite sure, is crucial to cardiac development in the early embryo. The electrical activity is present very early in life with resultant detectable contractions seen shortly after development of the neurotube. Would the heart develop if one could block the electrical signals? The architecture defining heart development and continued repair is not one of collagen, but is of one of electrical forces and electrical channels. The stem cells are surprisingly small to contain all of the information to become any cell or individual they choose. The stem cells are the bricks and the electrical forces are the brick layers. This concept of bricks and brick layers is my hypothesis of cardiac regeneration.

Often, patients with left bundle branch block will develop congestive heart failure that can be improved with resynchronization therapy by placement of an additional left ventricular pacemaker lead. In other instances, patients who are given right ventricular pacing leads and AV nodal ablation cutting the electrical pathways can develop heart failure that is attributed to electrical dys-synchrony. Restoration of the heart with a third left ventricular lead can allow patients to go from a gasping respiratory distress to a normal life. Dyssynchrony can be explained by imagining the heart is a V-8 engine with two spark plug wires pulled off there respective plugs. The additional left ventricular lead reattaches the spark plugs so the heart runs smoother and more efficient. The synchronization can demonstrate immediate improvement due to the improved timing of a coordinated contraction. There also appears to be new remodeling which repairs the previous damage from the cardiac dyssynchrony. This means new cells are laid down where no cells were before.

The Mayo Clinic is one of the finest institutions in the care of cardiac patients. Hypertrophic cardiomyopathy with obstruction is a condition where the intraventricular septum is hypertrophied, thickened, and forms an obstruction to the outflow of blood from the left ventricle. The anterior leaflet of the mitral valve sails like an airplane wing and obstructs the flow of blood leaving the heart during contractions. This results in a gradient across the outflow track and a leak of the mitral valve causing greater work of the heart and increased pressure in the left atrium. When simultaneously present, the two conditions can result in significant symptoms of congestive heart failure and pass out spells with exertion. The gold standard method of treatment of this condition is surgical resection of the bulging intraventricular septum, requiring surgically opening the chest and in good hands a 1% chance of mortality. In most cases this condition is a congenital defect, involving abnormal myocardial proteins and a disarray of muscular elements.

One possible non-surgical method of treatment of this condition is to place a pacemaker that will introduce cardiac dys-synchrony. By causing the septum to move in the opposite direction the outflow tract obstruction can be relieved. This has been demonstrated in the laboratory and has given some patients benefit. A clinical trial was initiated by the Mayo Clinic to prove or disprove this theory. A pacemaker was placed in the right ventricle in patients and the pacemaker was turned on or off in a blinded manner. The study demonstrated that there was no difference whether the pacemaker was functioning or not. RA Nishimura, JM Trusty, DL Hayes, DM Ilstrup, DR Larson, SN Hayes, TG Allison, and AJ *Tajik Dual-chamber pacing for hypertrophic cardiomyopathy: a randomized, double-blind, crossover trial* J Am Coll Cardiol, 1997; 29:435-441. The conclusions indicated that the act of putting in a pacemaker had a very strong placebo affect. The patients who were studied had the AV delay shortened to be sure that the pacemaker would capture the ventricle and cause cardiac dyssynchrony. Shortening the AV delay to maintain 100% pacemaker function causes less efficient filling of the left ventricle. This inefficiency can lead to continued symptoms due to inadequate filling. The electrical skeleton of the heart was not altered so remodeling could not occur.

If one believes in electrical remodeling, there would be no change in the architecture of the heart as long as the AV node was intact allowing electrical signals to pass through the normal fibers. For this reason, when placing a pacemaker for control of symptoms of hypertrophic cardiomyopathy with obstructions we have elected to ablate, radiofrequency burn, and eliminate the AV node to terminate the electrical pathways to the ventricle. The electrical pathways are now completely determined by the pacemaker and AV interval can be set that provides the most efficient pumping. A defibrillator is typically placed as part of the pacemaker to protect against sudden cardiac death. The condition of hypertrophic cardiomyopathy has a high propensity for sudden cardiac death. Observations indicate that the patients have improvement in their gradient initially and over the next five to six months there is continued remodeling of the heart and lessening of the intraventricular gradient. Severely symptomatic patients, already with a pacemaker for several years with an intact AV node have been rendered completely asymptomatic by ablating the AV node.

Some electrophysiologists feel that the AV node was constructed by God and should not be altered by man. For this reason ablation of the AV node is not popular. It results in the patient becoming completely reliant upon a pacemaker. Another less invasive approach in the patients with hypertrophic cardiomyopathy is to induce a myocardial infarction of the bulging septum by injecting alcohol in the first septal perforator. With this technique, the electrical conduction system may be damaged and pacemakers are required. Another benefit of the surgical procedures either ablative or cutting includes the destruction of that electrical pathway which will allow positive remodeling to occur with pacing.

If this concept is correct, rebuilding the heart with the help of electrical signals (brick layers) guiding stem cells (bricks) into the infarcted region may be a method of rebuilding broken hearts.

Murray's problem was that he had an interruption of the electrical skeleton of the heart with the condition of LBBB, left bundle branch block. There are two main trunks of the electrical wiring of the heart – the left and the right bundles. With LBBB the septum is activated late because the electrical impulse is interrupted and the heart contracts in a lopsided method losing efficiency and sometimes causing a greater mitral valve leak. The best analogy is a V-8 engine with two spark plug wires disconnected. By placing a third lead thru the venous system of the heart to the lateral wall the spark plug wires can be reconnected. Murray also received a defibrillator that would monitor his heart for rapid chaotic beats and would shock him back to life if a soul releasing rhythm persisted.

The old soldier trundled into the Houckster's office. Immediate symptom relief that other patients had told him he would experience never occurred. He was depressed. The Houckster listened to his chest and the soft S1 and paradoxical S2 persisted. The Houckster smiled, looked at the dejected Murray and said, "It is time for your tune-up." Murray grabbed his orders and walked defiantly down the hall to the echo lab where he encountered another beauty-the echo technician, a pacemaker representative and an eager young fellow in training. The echo tech made him happy she was at his side. She was slender with short hair and was ageless. She could have been 22 or 42 years of age. Her experience with the complex $250,000 echo machine was a give away for her older age. The General Electric machine saved hours when it came to resynchronization timing. The pacemaker representative would change the timing of the RV and LV leads and all would watch the screen for the green and red hues. Once the heart looked green, the RV and LV leads were properly timed. The AV interval was then adjusted to improve the filling. Finally, the LVOT velocity was remapped and confirmed that the changes had increased Murray's pumping capacity by 20%. Murray left the lab with more vigor in his step. Hope was rising again.

The next several sections of this chapter are more predictions and can not be completely backed by science. Predictions and modeling requires creativity. Creativity is a mystery and relates to our television shows that have the same plots from Roman theater. Without creativity or an open mind new discoveries will not follow. I want the student to think of predictions in a creative manner and not be bound by current thinking. Once the ideas are on paper experiments to prove or disapprove will follow. The following collections are not meant to believe but to think.

The Equation of Life

A teaching physician is more than an honorable profession and has great responsibility. A mentor is an individual who teaches his students lessons that were painful to learn the first time. Young doctors learn by making mistakes. Internship is an opportunity to make mistakes. The staff physician's job is to prevent those mistakes from harming patients. Shaping young minds and plying them with information that they would use in future patient encounters is an awesome responsibility. There are not many mentors who claimed to teach a mathematical equation of life.

Every cardiologist is familiar with the **Fick Equation** to calculate cardiac output. I would like to rewrite this equation and call it the **Equation of Life**.

Equation of Life

$$CO \times Hgb = O_2 \text{ consumption} / (PVO_2 \text{ Sat} - PAO_2 \text{ Sat}) 1.36$$

The terms on the right have relatively fixed values for humans. This equation can be used to calculate the lowest haemoglobin (Hgb) that is possible for *life* when the cardiac output (CO) is maximal. It also demonstrates that if your cardiac output is fixed at a low level, *life* can be maintained by increasing the Hgb. Further studies to prove a physical law are unnecessary. The physical examination findings for the equation of life are the assessments of anemia. By careful study, a physician can learn to guess the haemoglobin within one half gram by looking at the palpaebrae of the eye. This is the pink portion that lies behind the lower eyelid. Estimating how pink the palpaebrae surface can give a pretty good estimate of the haemoglobin or the perfusion of the patient. There are always false positives that occur due to conjunctiva injection from allergies. Poor perfusion will demonstrate a very pale palpaebrae. Visualization and estimation of this mucosal bed can determine the equation of life. Safety studies to determine the best method of raising the Hgb in our patients are necessary. Paying for the therapy by insurance companies and the government is essential.

Many heart failure doctors are already utilizing erythropoietin to improve symptoms using the Fick equation as justification. The symptomatic improvement in the individual patient is the positive feedback that these physicians use to justify use of this medication. There have been some recent studies that have questioned whether this is a valid practice or not.

Initial concerns utilizing blood transfusions came in studies of acute coronary syndrome. It was found that patients who received blood transfusions did worse than those who did not receive blood transfusions. The studies were retrospective and the conclusions that were drawn are that giving blood transfusions caused worst outcome. Patients who got blood transfusions were undoubtedly sicker than the patients that did not receive blood transfusions. Bleeding and heart attacks is not a friendly combination. Heart attacks require intense blockade of the clotting system that often results in bleeding. This side effect of the intense anticoagulation has serious ramifications. When a person bleeds the clotting system is increased to help save the patient from blood loss. In a patient who has a predisposition to form blood clots in their blood vessels this is a bad scenario. Bleeding begets clotting that causes re-infarction. People who bleed during their heart attacks will have worse outcomes than those patients who do not bleed. The proper conclusion from this study is that bleeding is a bad outcome in patients who have heart attacks, not that blood transfusion in patients who have heart attacks is bad.

Early studies in congestive heart failure and anemia have demonstrated that anemia is a strong predictor for bad outcomes. Patients with heart failure have lower hemoglobin than those patients without heart failure. The reason for this is uncertain but is unlikely to be dilution due to excessive retention of fluid. Biotech companies that produce substances to improve hemoglobin have attempted to do studies in heart failure patients and in renal dialysis patients. It was assumed that if hemoglobin were increased that mortality would be improved. In renal dialysis patients there is no difference in mortality. There are less hospitalizations and less congestive heart failure in patients treated with erythropoietin. Quality of life also appears to be improved. There has not been a drop in dialysis patient's death rate with the institution of this therapy. Erythropoietin therapy is approved in patients with elevated creatinine and lower-than-expected erythropoietin levels. Similar findings and conclusions were made in small studies in heart failure patients. In some patients raising the hemoglobin above the normal level can cause worse outcome. The proper use of these hormones in heart failure patients is still to be determined. The identification of patients who will benefit from this therapy is patients who are suffering under their current medical regime and have hemoglobin less than 11. A trial of this therapy, although expensive, is justified with the patient who receives the medication deciding if it helpful.

Stem Cells
Cellular therapy will produce an explosion of possibilities. The greatest medical advance will be when we learn how an embryo differentiates cells and turns them into organs. The therapies generated from this knowledge will not involve the sacrifice of embryos to produce a cure. The cure will be retraining adult stem cells to replace damaged organs. Embryos will have a role in showing us how an organism orders development and signals the cells to change in specific ways. Humans do not have to be studied to find these secrets.

Jennifer was 23 years old with a college degree, a job, and an ex-boyfriend. She also had a predisposition to a bipolar disorder. She did not understand why she felt so hopeless. She was bright, engaging, and even had been invited to dinner by a co-worker who learned of her availability. Fending illness she declined dinner, went home, and took a bottle of Acetaminophen. She did not feel bad until the next day when she began to vomit. She arrived to the Emergency Room too late for rescue from her overdose. She was given mucomyst anyway and supported. She quickly fell into a hepatic coma. She was initially treated with a live hepatic membrane that functioned as her liver. The cells had been donated by genetically deficient pig and kept alive by a nutrient solution. Her blood was circulated across this membrane and toxins were removed. She regained consciousness and was interviewed by a psychiatrist, gastroenterologist, and a geneticist. She had her blood and bone marrow pheresed for collection of hepatic cell progenitors and rare neuro cell progenitors. Her liver was re-constructed by her own cells after they had undergone amplification on the bench. They were injected into a vein and would take four months to regenerate the damaged liver. She was supported for the next four months by the pig liver membrane. During the four months her DNA was mapped and several genetic abnormalities were noted. The neuro progenitors were infected with a DNA sequence that would increase the neurotransmitter in her brain that had resulted in her deep depression. These cells were also injected and directed to the limbic system of the brain by magnetic direction. The cells had small vacuoles implanted that were high in iron content. Jennifer left the hospital 4 months after admission from a fatal illness. She never felt better.

Jennifer's story is totally science fiction but represents how this therapy may be used in the future to fix a problem and the root cause of the problem. Repairing the heart with stem cells is a scientific reality and is currently under investigation. There are many secrets of cells, their development and potential. Nature cures naturally. Our role is to understand this natural tendency to cure and not get in the way.

Plasmapheresis
The farther we are removed from the controversies of the past, the closer we are likely to resurrect an idea that was once thought idiotic. Bloodletters are fighting back. Plasmapheresis is a sophisticated from of blood letting. Blood is removed by a venous canula and sent to a machine that spins the fluid separating out the various components by centrifugal force. The serum can be removed and replaced by similar solutions. The blood cells can be added back and returned to the body. In this way the evil humors in the blood can be removed. This tool is already used in treatment of autoimmune diseases such as myasthenia gravis, thrombotic thrombocytopenia, and Guillain-Barré. Another form of this machine removes cholesterol, and others remove white cells in leukemia. Plasmapheresis can be used to remove cytokines that promote inflammation and vascular damage. This property makes it very appealing in the treatment of myocardial infarction. Inflammation is thought to be a cause of myocardial infarction with the inflammatory cells releasing enzymes that cause degradation of the fibrous cap. Necrosis (death) of the myocardial cells is a stimulus for inflammation. The inflammatory cells kill the stem cells, so that, instead of recovered myocardium, the injured myocardium is replaced by scar.

Plasmapheresis after myocardial infarction is appealing in that it will reduce the inflammatory state that caused the heart attack. It will also decrease fibrinogen levels so a recurrent thrombosis is less likely. By removing the cytokines, the natural replacement of the sick cells by new cells can be promoted. For an example of how this might work I will refer to Chapter 12: Plasmapheresis for myocardial infarction.

Hypothermia for Brain Injury and Shock

Cooling patients who are victims of sudden cardiac death to 93 degrees for 48 hours favors brain recovery. This therapy is known as hypothermia. During this time the patients seem to stabilize. The arrhythmias become less frequent. Blood pressure stabilizes at a low level and every thing slows down. Chemical reactions, enzymatic reactions are temperature dependent. Apoptosis is cell death and it requires chemical reactions to proceed. Desperately sick patients will be cooled immediately even before stabilization. In this cooled state blood will be replaced, bacterial toxins and evil humors will be removed by plasmapheresis.

Exercise and personal responsibility primary prevention
Exercise will be proved to help all aspects of cardiovascular disease. The mechanisms will be multiple. Increasing circulating stem cells will be the greatest benefit of exercise. For the weak in body *Enhanced Extracorpeal Counter Pulsation EECP* will be the answer but will require purchase of a machine.

Anilise is a 68 year old female who has multiple medical problems. She never considered herself as part of her health problems. She liked to eat, smoke, and drink on the weekends. Sweating only occurred during sex and after the change she had not experienced sweat for 12 years. Her good life nearly came to a halt when she developed crushing substernal chest pain. Without delay she presented to the emergency room and was quickly triaged to the cardiac catheterization laboratory. She had triple vessel disease including a closed posterior-lateral branch. She went to surgery and upon discharge was assigned several new diseases, Diabetes Mellitus, Hypertension, coronary artery disease, obesity, gout, and congestive heart failure secondary to diastolic dysfunction. She was discharged with 8 new medications.

She was signed up for cardiac rehabilitation and met an angel in the physical therapist. Cindy the physical therapist was a motivator and an educator. She would sit and educate Anilise for hours. She instructed her in proper exercise, diet, and provided spiritual enrichment. Anilise was motivated to get better. She completed cardiac rehab and continued in a phase 3 programs. She did additional exercise and became a health conscious cook. Cigarettes and all the paraphernalia were discarded.

In three years she had lost 60 pounds and lost five diagnoses. She was exercising for 2 hours per day, one hour before breakfast and one hour before supper. Anilise now looked 48 and had great vitality.

(EECP) Enhanced Extracorpeal Counter Pulsation
Christopher had a spinal degeneration and had become wheelchair dependent. Muscles that were not utilized became atrophied. He had many urinary tract infections and began to experience skin breakdown. He was referred to EECP *Enhanced Extracorpeal Counter Pulsation*. There he met Doug with severe peripheral neuropathy and painful legs from diabetes, and Fred who had Parkinson disease and was unable to walk due to shuffling gait and tremors. The only legitimate patient in the EECP room was Edna who was there receiving therapy for angina. EECP was approved only for treatment of angina and heart failure. The government reimbursed Edna's therapy. The rest were paying out

of pocket. They encouraged each other, as the air filled bladders squeezed blood into the veins and retrograde into the arterioles. The bladders were similar to a trauma suit and would inflate in sequence from the calf, thigh and hips during the resting phase of the cardiac cycle. The retrograde flow would knock off stem cells nestled among the endothelial cells. They would enter the circulation and perform their magic of rejuvenation. The improvement was dramatic. Edna's angina resolved after 4 weeks. Fred's gait became unassisted but he still had tremor. He was extending his treatment for an additional 35 treatments. Doug's painful neuropathy was relieved and some strength had returned to Christopher's legs.

Human Genome and designer drug therapies

The twenty-one year old was sleepy with a slight hangover. His first legal drink was the previous night. David had at least one too many. The designated driver got him home safe and even set his alarm. Today was to plot his health for the next 100 years. He had graduated college and was working for a small computer company. He had $60,000 of school debt and was going to drop $150,000 dollars for genome mapping and analysis. A sample of his blood would be drawn and the white cells would give up their DNA to a robotic sequencer that would determine his genetic makeup. David felt nausea, and dizzy just thinking about the coming blood draw. In the past he had fainted when he visualized his own blood. As much as he hated to have the blood draw, he knew his life was dependent on the sickening prick of a needle. In all of the advances in medicine, why did they still have to draw blood through a needle?

The sample would identify any mutation or polymorphism. These genes would determine what future diseases and illness David could expect. His metabolism and enzyme systems that interacted with various drugs would be recorded. In the end he would receive a piece of paper that would be his personalized roadmap to health and likely cause of death.

David

Cardiovascular risk	*** High 94%***
Abdominal Aneurysm risk	** Moderate 50%**
	*** High 90%*** if smoking
Diabetes risk	* low * if body weight maintained below 70 Kg
Hypertension risk	*** High ***
Stroke risk	*** High ***
Prostrate Cancer Risk	*** High 100% ***
Colon Cancer Risk	*** High 90% ***
Pneumonia	* Low *
HIV	** Moderate 50% **

Enzyme pathways
    Cytochrome P450   Low Normal
    CYP2D6   Low Normal

Beta 2 receptor   abnormal

The above analysis implies a high risk of developing cardiovascular disease and colon cancer.

Life-style should include 90 minutes of exercise per day for a calorie output of 1500 calories. Smoking is prohibited. Weight must be maintained within 10% of ideal body weight. Fiber, fruits and vegetables should be consumed two times per day.

A statin medication should be started immediately Simvistatin 20 mg at night
At age 35 ACEI should be started (Lisinopril 10 mg at night) along with a beta-blocker (Metoprolol 50mg two times per day). Aspirin should be started at age 55 with the use of Prilosec.

Screening colonoscopy will begin at age 35 and continue life long for every 5 years.
MRA of the aorta and coronaries will start at age 50.
Prostate ablation is recommended at age 65.

The end of the report directed David to a lifestyle monitoring section to have subcutaneous chip inserted for exercise monitoring and metabolic monitoring. This chip will be used to locate you for your free medications and notify you of screening events. It will be used to monitor your adherence to lifestyle. You are healthier than 75% of the population and will have a natural life of 125 years if these instructions are followed. At age 120 you may be eligible for a telomere extension. End of Message.

Science fiction   **Sewage management savior and the great undoing**
In the nineteenth century, odor, disease, and contamination of water supplies led to the development of sewage removal and ultimately to sewage treatment. These civil systems have saved more lives than any advancement in medicine. Today the flush toilet is taken for granted and numerous items descend into the depths including DNA.

The bacteria and viruses, that are shed into this system survive the harsh treatments and can incorporate new DNA into its structure. Antibiotic resistance emerges and is shared with other bacteria. Could other DNA fragments be incorporated that would cause a new species to develop?

The bubbling gruel in the sewage treatment plant seemed unchanged and certainly did not have the appearance of a birthing room. Three E.Coli Bacterial membranes were slightly damaged and a SNP (snip) of DNA traveled within reach of the middle organism. It was the code for sodium channel. The little E. Coli picked up the strand and soon it was in the nucleus producing sodium channel proteins. The E. Coli divided and reproduced two daughter cells and within 24 hours there were 100,000 E.Coli with the new proteins. The E. Coli were unaware but the channels kept the sisters together. The new channels allowed communication between the sisters so finding food became much more efficient. The leaky membranes soon found other snippets of DNA and new properties arose from the now massive bacteria colony. Mobility and awareness arose and a new predator was on the block –The Blob.

Science Fiction   The Earth is alive

How can we define ourselves? What is life? Do we consider mitochondria alive? They have their own DNA, they reproduce and they seem to have purpose by supplying energy to maintain our cells. One definition of life is a system that has negative entropy. Negative entropy describes a system that becomes more ordered than disordered over time with the utilization of energy. Is the Earth becoming more ordered? Are we the mitochondria of the Earth?

Our planet was a place of chaos with volcanoes and putrid atmosphere. The Earth made a change and slowly the atmosphere changed, plants sprung upon the earth and stored energy as part of the carbon cycle. The energy was placed into the earth in the form of carbon so that it could be released at a later date. Energy was needed to counter chaos. Humans sprang forth and had the audacity to believe they were the great life force and were unaware that the planet was merely using them as a tool.

2100 - The great nuclear engines on the Moon had been in operation for 50 years. The moon was now spinning at slightly less than one rotation per day. The products of the nuclear engines that had propelled the Moon were water vapor, carbon, and silica. Plants were now flourishing in shelters and would soon be placed under the new atmosphere of the moon. The moon was becoming more ordered. Graviton waves emanated from the Earth and showered the Moon tenderly. Instantaneously, the entire universe knew of the birth. Earth was a mother. The Moon was alive. Humans were unaware of the communication. They had no means of understanding gravity and the strings that tied the entire Universe together. The concept of a higher intellect was not in their grasp even when the intellect was under their feet.

Heart Failure

Heart failure is mysterious. The more we learn about the mysteries of heart failure the more complicated and profound they become. Why are heart failure patients anemic? Why does blood pressure plummet in stable patients when a cause cannot be found? The "Ying and the Yang" philosophy is ever present in life. All biological systems are in a balance and a dip in one direction can cause a disease state. In some cases measurement of these hormones cannot find the dip. Recently a paper suggested that elevated BNP in heart failure patients was a failure to cleave the pro molecule to the active form. The measurement technique could not find a difference between these two forms. To the physician the "Ying and Yang" elevated neurohormones and its counter BNP were matched. The cells in the body, however, did not recognize the un-cleaved BNP and the neurohormones dipped the patient toward heart failure. This concept of failure to cleave an active molecule may explain many disease conditions. What could be interfering with the molecular cleavage? The answer could be lack of enzyme, turning off an enzyme due to physiological milieu, toxin blockers, bad Karma. These new questions need to be answered.

These questions are the future, but the past lessons need to be remembered. The next two chapters are thinking chapters and slip centuries into the past and back to the present. The art of physical exam has been passed from physician to physician and it is to those mentors I give tribute. Chapter 10 deals with physical examination, a poorly performed art that has fallen to the impersonal technology. Chapter 11 explains hemodynamics, or how medications along with a good physical exam can improve symptoms. If you have a mind willing to learn or if you are a practicing physician who wants to get back to the basics proceed. Otherwise skip to chapter 12.

Chapter 10
Physical Exam – The Way It Should Be Taught

The Houckster walked into the auditorium filled with residents and medical students who were captive as part of a required lecture series. Residents' time was precious, so precious they were educated even during their lunch hour. Training to work the lunch hour was another subliminal message to be an efficient health care provider. The more efficient the health care provider, the more money the health care industry can make. The fast pace of medicine demands every second of the day be accounted. Therefore, it was considered normal to listen to a lecture while chomping on a sandwich. The Houckster strolled to the podium and set up his trusty PowerPoint presentation to help illustrate the proper way of doing physical examination. He looked to the back row to see what residents he could harass that were obviously attempting to escape his inquiries.

The residents were interested but intimidated in physical examination. They found it difficult and had a secret hope that in today's world it was not important. The physical exam required more than book knowledge. It required sensory perception. The accuracy of the exam paled in comparison with a detailed picture of the heart, vascular system that could be obtained with expensive techniques such as echocardiography, CT scan imaging of the heart, and MRI of the heart. The second mistake the residents and students would make was to trust the interpretation of these images by other physicians. They were breaking the first rule of the Houckster trusting someone else's opinion: "Do not trust your mother." The students, residents, and fellows all lacked confidence in their examinations and were reluctant to commit therapy based on their senses. Losses of the physical examination skills have paralleled the advance of technology. The more the young physicians would rely on technology, the worse their examination. The time that physical examination was most important was when there was no time for technology. Physical exam is faster than any ordered test.

Some of the medicine residents and students would do mission work during vacations in far away lands. In these faraway lands where technology was not available, physicians demonstrate the importance of physical examination when all they had to rely on was their own senses. Technology was not available to bail them out of a lazy physical examination. Physical examination was learned by necessity. Most students don't have that luxury of going to the Third World to hone their physical examination skills.

Some academic centers have argued that the physical examination is a lost art and will be replaced by technology. Handheld echocardiograms may replace the stethoscope that sits in one's pocket. The instructors who are claiming to look into the future may be surprised by its failure. Let's examine a simple technology introduced over 100 years ago. The electrocardiogram has only 12 leads and six waves. This piece of paper is one of the most misdiagnosed documents and is simple compared to the echocardiogram. An echocardiogram shows a heart in two dimensions although it is a three-dimensional structure. The image is moving in time and demonstrates flows within the heart with numerous artifacts that can be imposed upon the images to confuse the reader. I have no hope that the general physician can read an echocardiogram since most general physicians fail at reading a 12 lead EKG.

The student is not entirely to blame. The teaching in medical schools is often performed by physiologists and is not taught in a proper manner. While methodology seems practical and all the texts teach the same thing, most human brains do not function in that manner. The easiest way for the brain to work is to deal with a limited number of possibilities and eliminate implausible possibilities. This way of thinking can be formally modeled as Bayes' Theorem. Bayes' Theorem tells how to update or revise beliefs in light of new evidence (**Wikipedia**). Instead, medical students are taught to be a recorder of the senses and are not expected to think while recording. The student diagrams heart sounds using several slashes and squiggles to represent the sound that he hears. These symbols are compared to certain abnormal cardiac findings in the textbooks and a student should then be able to pick out the abnormality. After recording these cryptic symbols the medical student has to make a diagnosis of the heart findings that caused this particular type of murmur. Physical examination is impossible to do in this manner. A soft sound can not be heard unless you are expecting to hear it. It will be quite unfortunate if the stethoscope is relegated to the position of the slide rule collecting dust in an ancient closet. The physical examination is always faster than any piece of technology.

Approach to the
Cardiac Physical Exam
Rule 1

Engage Brain before Senses

Like Sherlock Holmes, a physician uses clues to develop the differential diagnosis eliminating certain conditions and honing in others based on the clues that are presented. The first rule of physical examination is to engage the brain before the senses. The brain begins to process information as the patient walks into the room. The brain will register the general appearance of the patient, the patient's pallor, and speaking voice. Illness and the severity of discomfort will become quite evident. The brain then asks questions to determine the symptom complex of the patient in an effort to eliminate some diagnoses and confirm others. The manner in which questions are asked is critical to obtain correct information.

For example, if you came across a farmer with several missing fingers from previous accidents and asked if he was experiencing chest pain, he would most likely respond with a "no". Several hours later when the staff physician enquires if he had discomfort or any unusual discomfort in the chest, he will reply, "yes". Collecting information from the patient's perspective is critical in making a diagnosis. The ten dangerous conditions with similar complaints can be narrowed and eliminated saving many dollars in tests and getting the answer as quickly as possible. The physical examination is used with the history by the brain to either confirm or reject a diagnosis. The physical exam must be completed in an organized rigid sequence.

The physician always examines the patient from the patient's right side. The fingers can then properly feel the size of the organs and pulsations of blood vessels. The exam always begins with the vital signs and blood pressure in both arms. The blood count is estimated by observing the pink of the palpaebrae. The neck is observed for jugular venous distension while the patient is seated. Scalp veins, pallor, hair patterns, ear lobe creases and skin condition are all observed in 10 seconds or less. The veins are used to measure the filling pressure of the right heart. The pulsations take even greater skill. X descent and Y descent can be very important. After the exam is concluded and raises the suspicion of tamponade or constrictive pericarditis, the veins can be re-examined. X descent favors tamponade, Y descent favors constriction.

Rule 2

The Exam is in the fingertips

Use palpation to determine:

- **The state of the *systemic circulation***
- **The state of the *pulmonary circulation***
- **The state of the *left ventricle***

The above information is used to determine the presence of valvular and cardiac abnormalities.
Auscultate to confirm diagnosis and severity of the valvular dysfunction.

The state of the ***systemic circulation*** is determined by the carotid exam. Feeling adjacent to the voice box in the groove of the neck the carotid artery is found. The carotid is one of the first major branches off of the aorta and its pulsation is a reflection of the output of the heart. Placing two to three fingers along the length of the vessel will give rise to a pulsed sensation. The examiner will determine the volume of the pulse, if it is weak or bounding, delayed in upstroke. Occasionally, a vibratory thrill, a spike and dome pattern, or a notched pattern will signal a diagnosis. The main objective is to decide if the pulse is normal, hyperdynamic or a low volume pulse. For each of these states, normal, decreased and hyperdynamic there will be a differential diagnosis that will eventually point to a diagnosis.

Increased Carotid Pulsation – Hyperdynamic pulsations are associated with Aortic Insufficiency and Mitral Regurgitation
        Systemic conditions that increase Cardiac Output:
           •Anemia
           •Pregnancy
           •Infection
           •Thyrotoxicosis
           •Peripheral Shunts

Anemia, pregnancy, infection, thyrotoxicosis, and peripheral shunts are all due to increased cardiac output. Ann was a 30 year-old female who was referred to the cardiologist for a new murmur. The cardiologist palpated her carotid artery and noted that it was quite hyperdynamic and mildly tachycardia. He estimated her heart rate to be just under 100 beats per minute. He listened to her heart and heard only a pulmonary flow murmur that was quite soft and no diastolic murmurs. He again looked at her cheeks and found her to have a rosy light brown patchy complexion and casually asked her when was her last menstrual cycle? She stated her last menstrual period was between six weeks and two months ago. The cardiologist then inquired as to whether she may be pregnant. She stated that they've been trying for several years and she was scheduled that very afternoon to drop off a urine sample to help confirm her pregnancy. The cardiologist told her that based on her carotid exam she was pregnant. Eight months later a new baby boy was brought into the world. The carotid upstroke was quite dynamic, because even in early pregnancy cardiac output increases to help to provide more nutrients to the developing embryo.

Frieda was visiting the cardiologist as a routine follow up for her atrial fibrillation. She had been feeling more fatigue and had a desire to chomp on ice throughout the day. During her examination the cardiologist noted palpaebrae to be quite pale and upon feeling her carotid artery he noted a very dynamic pulse. The heart rate had also increased from her normal 70 beats per minute to close to 100 beats per minute. He was saddened by these findings because he knew that iron deficiency anemia was the most likely reason for this finding. In her age group colon cancer was on top of the list. Colonoscopy was performed and demonstrated in 5 cm mass in the ascending colon.

Aortic insufficiency and mitral regurgitation are two valvular abnormalities that both have a hyperdynamic carotid feel for two different reasons. Aortic insufficiency is condition describing a leaky aortic valve which is the main valve leaving the heart to the body. Five liters of forward flow is necessary to supply the needs of the average adult. If two liters fall back into the ventricle because of the leak, the heart has to pump the original required 5 liters plus the additional 2 liters for a total of 7 liters of blood. This volume of blood traveling towards the carotid artery will give it a very brisk and hyper-dynamic feel. If the aortic insufficiency is quite severe the dicrotic notch which correlates with aortic valve closure can be palpated.

Eugene, a retired banker had been to his dentist (frequently and incorrectly blamed as the cause of endocarditis) six weeks prior and now was experiencing daily chills and low grade fevers. He noted black streaks on his nails and several tender spots on his palm. His primary care provider heard a systolic murmur and referred him to cardiology. Splinter hemorrhages, Osler's nodes, and Janeway lesions had been obvious. The carotids were bounding. Initial auscultation revealed a loud systolic III-IV out of VI systolic murmur. The murmur was so loud it was difficult to concentrate on diastole where a soft I/VI murmur was auscultated. Blood cultures were obtained and were all positive. Echocardiogram revealed aortic vegetation that was 1 cm in diameter with torrential aortic insufficiency. Twenty-four hours of antibiotics were given when the valve collapsed and Eugene was rushed to emergency valve replacement for his infected aortic valve. Eugene eventual returned to the human race after a lengthy hospitalization that included a small stroke and six weeks of antibiotics.

The reason mitral regurgitation is associated with a dynamic carotid upstroke is a little bit more difficult to understand. The mitral valve is between the main pumping chamber of the heart and the left atrium which collects blood from the lungs. When the mitral valve becomes leaky it is much easier for the heart to empty because it empties both into the aorta (high resistance) and into the left atrium (low resistance). The ejection fraction (the percentage of volume that leaves the heart) increases from 60 to 70% or greater. This increased ejection fraction results in a rapid upstroke and fall of blood in the carotid, resulting in a vigorous pulse in the carotid. Even when the ventricle begins to fail from mitral regurgitation there is early dynamic rise of the contour of the carotid with a later rapid falloff.

Judy S. was a high school junior who was trying out for the volleyball team. She had grown 6 inches in the last year and was a commanding 6 ft 2 inches. Despite her rapid growth she was able to command her body and was a good athlete. During her sports physical a loud murmur was heard and Judy was referred to the cardiologist for athletic participation. The murmur was loud and heard best anteriorly over the right chest wall with components of the murmur radiating to the neck and throughout the chest wall. The murmur in the neck is consistent with aortic stenosis and a 50 mmHg gradient would prohibit her from participation. The carotids, however, had a super dynamic feel so aortic stenosis was eliminated. Hand grip caused an increase of the murmur and standing made the murmur longer. If aortic stenosis was present the murmur should become softer with handgrip. The handgrip maneuver increases peripheral vascular resistance and decreases cardiac output reducing the flow across the aortic valve. Reduction of flow results in a softer murmur. The diagnosis was mitral valve prolapse and was confirmed by having the patient squat and listening for the clicks to move further into systole. Judy's arm length was greater than height confirming a diagnosis of Marfan's syndrome. Beta blockers were prescribed and the aortic root was determined to be normal in size with 2-dimensional echocardiography. Judy was also enrolled in a study to see if Losartan would prevent the dilatation of the aorta.

When the carotid upstroke is hyperdynamic, only two valvular abnormalities are considered: aortic insufficiency and mitral regurgitation. Other systemic reasons for this finding are due to disease processes that increase the cardiac output: anemia, shunts, infection, pregnancy, and hyperthyroidism.

Decreased Carotid Pulsation – low flow is associated with:
> •**Aortic, Mitral, Pulmonary Stenosis**
> •**Sick Heart Muscle, CHF**
> •**Pulmonary Hypertension**

Any stenosis within the cardiovascular circuit will reduce the blood flow. Decreased cardiac output is the explanation for low volume carotid upstroke in the valvular abnormalities of aortic, mitral, and pulmonary stenosis. Aortic stenosis is the most common and has a delay in the upstroke because the blood is struggling to get out of the heart through the calcified poorly compliant valve. If the gradient is more than 50 mmHg there can be a vibratory feel known as a 'thrill' in the neck. The diagnosis is then easy to make.

I have taught fellows that if the carotid has a poor contour that aortic stenosis must be ruled out. One of my Air Force fellows was asked to see an old retired Marine who was on the ventilator due to a COPD exacerbation. He was experiencing frequent exacerbations of his lung disease. The fellow was consulted because of some extra heart beats. He did his exam and wrote a note and went home. He was lying in bed and was recounting his exam and thought it was peculiar that he could not feel a very good carotid pulse. The patient had a barrel chest and had no murmur of aortic stenosis. He knew what the Houckster expected. Reluctantly, he got himself out of bed at 0100 and returned to the hospital to perform an echocardiogram to rule out aortic stenosis. The echo was a surprise and demonstrated a rock hard non-mobile aortic valve and a left ventricle that was dilated and very weak. To have a murmur there has to be a gradient. For a gradient there must be flow. The heart had gotten so weak the flow was very poor. The next morning the Houckster confirmed there was no murmur. The poor Marine never had COPD but had heart failure due to critical aortic stenosis. Although the surgery was risky, he completely returned to an active life and never had a COPD exacerbation after this procedure.

If the heart is weak due to a sick heart muscle or previous infarction, the cardiac output will be low since the pumping function of the heart is compromised. When feeling the carotid, heart failure is the most common reason for a poor carotid upstroke. With better medications longevity has improved in this terrible disease accelerating the number of people with heart failure. The carotid upstroke of these individuals is low volume because the cardiac output is low. In addition, the skin is often cool and clammy. The patients are not pleasant to touch. Feeling the carotids and observing the frequent shallow breaths is all that is needed to suggest a proper course of therapy.

Pulmonary hypertension, a diagnosis similar to mitral stenosis creates a resistance to flow across the lungs and reduces cardiac output. Once rare and affecting only 2/100,000 patients, the frequency of this condition is increasing and can be deadly if not properly diagnosed and treated. The current incidence of primary pulmonary hypertension is not known, but the secondary causes have risen due to obesity and the sleep apnea that accompanies obesity. Struggling for air during sleep raises the pulmonary pressures.

The state of the ***pulmonary circulation*** is determined by a very important physical exam finding. The finding is subtle and takes skill to elicit. The fingers are placed into the 2nd 3rd and 4th intercostals spaces adjacent to the left sternum. If you remove the skin, cartilage of the chest wall your fingers would be touching the pulmonary artery. During expiration, when the heart is closest to the chest wall, a soft pulsation can be elicited in patients with pulmonary hypertension. The pulmonary artery usually has a low pressure of 25 mmHg and this low pressure will not be transmitted through the rigid chest. In diseased states the pressure will double or triple. At about 50 mmHg, the pulsation in the pulmonary artery is strong enough to pass through the chest wall to your delicate hypersensitive fingers. As the pulmonary pressure rises, the right heart becomes more hypertrophied and larger. Eventually the right heart will lift the sternum and this is known as a right heart heave. The earliest finding of pulmonary hypertension is a pulmonary tap **(Pulmonary Systolic Pulsation)** during expiration. The right heart heave is a late finding.

Pulmonary Systolic Pulsation (+) Tap Present
   - **Mitral valve Disease - Mitral Stenosis or Mitral Regurgitation**
   - **CHF**
   - **Pulmonary Hypertension**
   - **Atrial Septal Defect**
   - **Young Thin Patient**

There can be false positives in any test. False positive means your test is positive but the disease is not present. A young thin chest wall can transmit a pulsation to the chest wall especially if the young thin patient is pregnant and has three times as much blood traveling in the pulmonary artery due to the pregnancy.

The pulmonary artery can be palpated in a patient with an atrial septal defect because 2 to 3 times the amount of blood is traveling in the pulmonary artery due to a shunt of blood from the left atrium that passes into the right atrium. The shunt causes the right heart to pump more blood than the left heart. The extra flow will raise the pressure in the lungs. If this goes on too long the pressure elevation becomes non-reversible. A patent foramen ovale (PFO) is a tiny shunt that occurs in 25% of all patients. During fetal development there is no need to pump blood to the lungs. The PFO is the bypass and closes soon after birth. The low pressure of the lungs allows the membrane to close off the flow. In 25% of individuals there is still a small communication. This becomes important in paradoxical stroke when a blood clot can pass from the right atrium across this communication to the brain. It is also important in skin divers who may suffer more brain damage from the bends. Patients with PFO's tend to have more migraines and strokes than those without this finding.

Pulmonary hypertension is an elevation in pulmonary pressure because the resistance to flow is high. The lung vasculature bifurcates many times so the resistance to flow is minimal. The cross sectional area increases as you go more distal. There is a very low pressure drop across the lungs. In pulmonary hypertension the cross sectional area decreases due to the muscular arterioles becoming thick and hypertrophied. Since the cross sectional area is smaller the flow is through smaller vessels requiring a greater pressure gradient. The condition is serious and most patients with this disease will be dead within 5 years. The increased pressure will be transmitted through the pulmonary artery to the chest and can be felt as a tap. This may be the only abnormal finding.

Betty was a 25 year old mother of 2 young boys. She had noted increased fatigue and shortness of breath. Her symptoms had progressed over the last three months. The patient went to her primary care provider who could not find anything obvious. She had last delivered 6 months ago and he felt the likely diagnosis was depression and started an anti-depressant. In three months there was no improvement and she was referred to psychiatry for further therapy. Prior to electric shock therapy an EKG was abnormal and she was referred to cardiology. The cardiologist felt a prominent systolic pulsation of the pulmonary artery. An echocardiogram demonstrated a hypertrophied dilated right ventricle with a small under filled left heart. Tricuspid regurgitation demonstrated a continuous wave doppler signal of 4.5 m/sec. This is equivalent to a right heart pressure of nearly 100 mmHg. The poor young woman was not depressed but had a deadly disease. She would have been referred to pulmonary for this deadly condition if the pulmonary tap had been detected.

Decompensated Heart failure, and significant mitral valve disease such as stenosis or regurgitation will elevate left atrial pressure. The right heart pressure and pulmonary pressures have to increase to remain higher than the left atrial pressure. If the right heart pressures did not become elevated blood would flow backwards. The pressure elevation from mitral valve disease or from congestive heart failure has a bad prognosis. Feeling the pulmonary tap is a cheap simple method to determine if a patient needs surgery. The reason to do surgery in mitral valve disease correlates very well with elevated right heart pressures over 50 to 55 mmHg.

Andrew was being followed for mitral valve prolapse. He had this diagnosed 20 years prior and was now 65 years of age. The mitral valve leak was the reason he was being seen on an annual basis. He never had complaints but was beginning to give up some activity he attributed to just getting older. His exam today demonstrated a pulmonary tap. Review of previous examination did not demonstrate this finding. Echocardiogram revealed a chord was torn with elevation of right heart pressures. Andrew was referred to surgery for a repair of the posterior leaflet, mitral valve ring. He recovered and was back to doing all of his activities.

Another example of how important the systolic pulsation of the pulmonary artery comes from The Houckster's previous experience while stationed in the Philippines and sitting in a remote mobile medical hospital. The desert tents were part of the third Air Force stationed in the Philippines. The tents were designed to collect water and were not very practical in the tropical Philippines. Their intended target was more than 8,000 miles away in Saudi Arabia. The exercise had been to deploy the hospital and train the troops in rapid deployment of the hospital. It was fun watching the 4 ton pallets fall out of the sky from C-130's until you realized they were falling into your space. After the hospital was deployed and a free clinic set up as a training event three thousand natives lined up for vitamins, antibiotics, dental work, and minor surgery. The volume of patients required screening examination to be less than 35 seconds. After 35 seconds a diagnosis would be made and vitamins, antibiotics, or a sad pat on the back would be made.

One of these patients was a 17 year old who was asthmatic and experiencing some wheezing. The Airman accompanying the young man told the Houckster that the man had asthma and inquired if she should take him to the emergency room. Initially the Houckster said yes. In the emergency room he would be given epinephrine for his asthma. Remembering not to trust his mother, the first rule of the Houckster, he recanted that order and asked the youth to be brought to the front of the line so a physical exam could be performed to confirm the diagnosis. His exam consisted of feeling the carotid which was irregular and of poor volume, and palpating the pulmonary artery which had a prominent tap. A shaky feeling had fallen over the Houckster. He had nearly killed the youth with epinephrine. Mitral stenosis was obvious. The youth was in florid pulmonary edema and had cardiac asthma and not pulmonary asthma. Digoxin and Lasix were prescribed with a least a brief improvement in the patient's symptoms.

The state of the **_left ventricle_** can be determined by feeling the apical impulse or PMI (the point of maximum impulse). Medical school wrongly taught the student to feel for the impulse with the patient lying on the left side – the left lateral decubitus. It was true the apex can be better felt because the heart is closer to the chest wall in this position. However, performing this maneuver in the lateral decubitus detracts from your sense of the size of the heart. The proper position should be in the seated position with the heart hanging in the chest at its normal position. The examiner should form an opinion as to the size of the heart and the contractile strength. An estimate of the ejection fraction should be attempted.

In the evenings when the echo machine is available, the order of business is history, physical, and then bedside echocardiogram. The physical exam and estimate of ejection fraction has immediate feedback from the echocardiogram. By following this procedure thousands of times, the echo can be predicted from a careful physical examination. The physical exam is improved by the echo.

Frank was a 55 year-old alcoholic who had congestive heart failure from the toxicity of excess alcohol. He had just been discharged from a lengthy hospital stay for decompensated congestive heart failure. He had been on a ventilator for four days and had nearly died. The experience was still very fresh in his mind even though the horrible event had occurred six month earlier. With the help of AA he had been sober since leaving the hospital. The abstinence from alcohol and the good work of Tracie in the heart failure clinic had repaired his heart. A new decision had to be made. If he went back to work he would lose his disability and perhaps place his family in financial jeopardy. He remembered how difficult it was to get the help that he now enjoyed. It had taken three months to be granted disability and he could not afford another three months without some kind of income. He had already gone through his retirement savings. He still had a child in school and did not wish to jeopardize her future. Frank decided to risk giving up his disability and go back to work. The Houckster told him that people on disability never did as well as those who were not. Frank is the exception and most individuals never return to work and their health suffers. The current system does not allow for people to better themselves and sadly promotes disease states.

He was due for his return visit to Dr. Houck. He offered no complaints and tolerated his medications well with the exception of getting dizzy when he would stand too quickly. His carotid upstroke was brisk. The pulmonary artery was not detectable. The point of maximum impact was not displaced small and discrete. The Houckster's estimate of ejection fraction was normal at 60%. The Houckster told Frank that he felt his exam was normal and he would order an echocardiogram to confirm his findings. He emphasized that return to alcohol would bring a rapid return of his heart dysfunction and that he should never imbibe again. Frank stayed on the wagon and proudly watched his daughter graduate from medical school. It was his illness and recovery that motivated her to go into medicine.

Determine the size and contractile strength of the ventricle (PMI) with the patient sitting up.

Small - **Mitral Stenosis, Pulmonary Hypertension**
Displaced, Discrete - **LVH, Aortic Stenosis (pressure overload)**
Displaced, Diffuse - **CHF, AI, (late) MR (late) (volume overload)**

If the ventricle can not be palpated the patient could be normal or have a small heart due to mitral stenosis or pulmonary hypertension. Statistical probability would vote for a normal patient. Some patients may be so large that the PMI can not be palpated or you are concentrating on the fungus in the skin fold that your hand has slipped and you can not concentrate on palpation.

A hypertrophied ventricle from aortic stenosis or hypertension (pressure overload) is like a fist hitting the wall. An estimate of the size and ejection fraction can be made by feeling how large and forceful the impact to the chest wall. The more lateral the PMI, the larger the heart and lower the ejection fraction. The softer the impact (less discrete) implies a smaller ejection fraction. Aortic stenosis and hypertension make the strongest impact. These ventricles over time will begin to fail so the enlarged discrete impact becomes more diffuse. In late stage aortic insufficiency and mitral regurgitation, the ventricle will also fail. The regurgitant lesions (volume overload) have a transition from normal to abnormal. A special form of the PMI can be palpated when the apex has been infarcted and is an aneurysm. The PMI is palpated when the heart fills and bounces against the chest wall. This occurs during diastole. If the impact is felt as the carotids are rising this is during systole and is due to the bulging of the aneurysm. The summary of the most common findings is below. Ear mark this page so it can be easily re-opened.

Increased Carotid Pulsation	**Pulmonary Artery Systolic Tap (+)**
Aortic Insufficiency	Mitral Valve Disease (MR or MS)
Mitral Regurgitation	CHF
Increased Cardiac Output	Pulmonary Hypertension

Pulmonary Artery Systolic Tap (-)
Normal Pulmonary Artery Pressures

Decreased Carotid Pulsation	**PMI sitting up**
Aortic Stenosis	Small - Mitral Stenosis,
Mitral Stenosis	Pulmonary Hypertension
Pulmonary Hypertension	Displaced, Discrete – LVH (htn.)
CHF sick heart muscle	Aortic stenosis
	Displaced, Diffuse - CHF
	AI, MR (late)

Carotid Pulsation	Pulmonary Tap	PMI

By palpating the carotid, pulmonary artery, and the PMI, the valvular status and systolic function can be determined. Auscultation can then be directed to finding pathology. It is more productive to listen intently for a sound you think is present than to simply listen. The real secret to auscultation is to know a sound is present. Example of how palpation can determine if a sound is present will follow a brief discussion of auscultation.

Auscultation

Auscultation was not always part of the physical exam. Observation of pallor, examining the jugular venous distention, and palpation of carotids and organs existed centuries before Dr. Lannec invented the stethoscope. This name may sound familiar as Dr. Lannec is better known to medical students in the diagnosis "Lannec's Cirrhosis" the late stage of alcoholic liver disease.

Placing your ear upon the chest was successful in hearing the tones and melodies of the heart. Certain murmurs, like the murmur of ventricular septal defect could cause a thrill a vibration on the chest wall. Placing an ear over this thrill would produce a harsh sound that was not supposed to be in the chest. If a physician of the court placed his ear next to the breast of a Lady of the Court, he may lose his ear or other valuable appendages. Lannec rolled up a piece of paper into a tube that could be placed between the clothing and was able to auscultate discretely without fear of losing his ear. A piece of paper can not be marketed so a wooden cylinder with an ear piece was developed. Lannec was the inventor of the stethoscope.

The principles of sound transmission have not changed. A short thick walled transmission tube is still best for capture of low frequency sounds. The sounds travel through the wall of the tube and only slightly through the air space. In some patients it would be preferable to be farther away, but the sound transmission degrades if the tube is too long. The head of the stethoscope has changed and is responsible for collected sound and filtering the sound. I prefer a Triple Head Proctor Harvey. There are three heads that swivel on a sturdy collecting chamber. One head is flat and will filter low frequency sound. It is known as a diaphragm. The diaphragm is a stiff thin piece of plastic that will vibrate and transmit sound to the collecting chamber. The diaphragm does the bulk of the work. Auscultation of low frequencies requires a bell, the second head that is a spherical shape and can be placed lightly on the skin. For the bell to work a seal must form at the skin. The low frequencies are then transmitted from the chest to the bell. The bell amplifies the low frequency. By pressing on the bell the skin is stretched and the stretched skin now acts like a diaphragm. If the sound becomes extinguished it confirms it was present. This is very useful is confirming if an S3 or S4 is present. The third head is rarely needed. It is a ridged diaphragm with raised circles that protrude from the center to the circumference. This sound quality of this head is in between a diaphragm and a bell. By only having contact with the small center circle the low frequencies are enhanced. Pressing harder will bring larger circles to the chest and the frequency transmitted will be higher. It is very valuable in children or boney patients. If the target of skin is small the little circle is great for fitting into this area even though the triple head is bulky. Neonates and small infants require special care. If too much pressure is placed on the chest the heart will slow down. It is often good to look at the monitor while auscultating so that your desire to hear better does not harm the infant. The same is true during an echocardiograhic examination.

The art of auscultation involves listening to murmurs (systole, diastole, continuous), heart sounds (S1, S2, S3, S4), clicks, opening snaps, honks and whoops, plops. There are some extra cardiac sounds such as rub which is very useful in diagnosing chest pain due to pericarditis.

So what makes the sound of a murmur. If you are lying next to a stream and can hear the soothing ripple of noises as the water passes over rocks, you are listening to a murmur - the murmur of the creek. If you were by a large stream that was flowing slowly you may not hear anything. As the water approaches a narrow gorge the flow accelerates becomes turbulent and the noise can become deafening especially if you are lying next to Niagara Falls. Murmurs are caused by turbulence and turbulence is caused by pressure gradients the pressure difference between chambers. Displayed are some graphical representations of cardiac murmurs demonstrating the gradients and why they are soft or loud. A soft murmur requires careful attention so that it will not be overlooked. Aortic stenosis and mitral regurgitation are both easily heard murmurs. Mitral stenosis and aortic insufficiency are soft and require intuitive suspicion to detect.

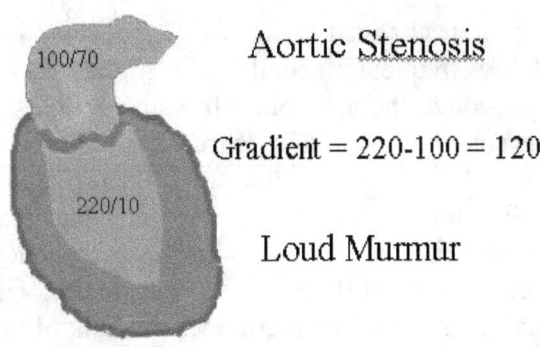

Aortic Stenosis

Gradient = 220-100 = 120

Loud Murmur

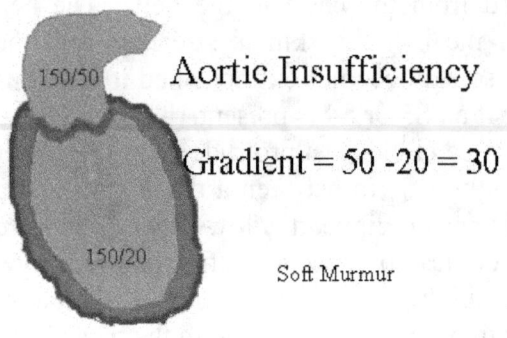

Aortic Insufficiency

Gradient = 50 -20 = 30

Soft Murmur

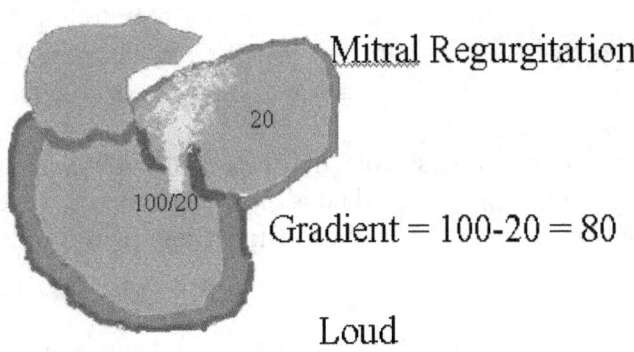

Mitral Regurgitation

Gradient = 100-20 = 80

Loud

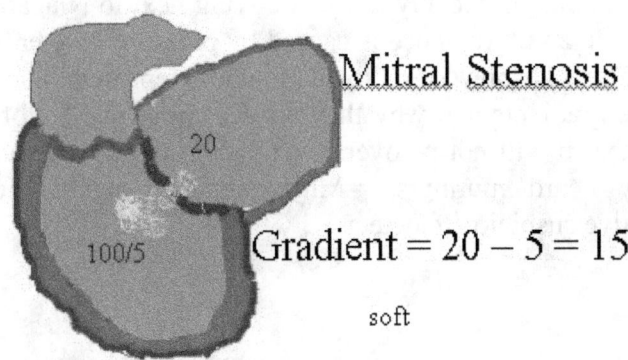

Mitral Stenosis

Gradient = 20 – 5 = 15

soft

Murmurs are classified as loud or soft, and are graded I through VI in intensity. I through IV distinguish the barely perceivable to the very loud. A V has an associated thrill and VI is when the murmur can be heard without the stethoscope being placed on the chest. The murmurs are also timed to systole the ejection time of the heart and diastole the filling time of the heart. The beginning student may have trouble distinguishing systole from diastole. Diastolic murmurs are almost always soft. To get the timing correct feel the carotid. The carotid pulsation occurs during ejection (systole). If you hear something when the carotid is not felt it qualifies as diastole. If you can not decide, you should always bet on systole.

There are three conditions when a murmur is continuously heard through out the cardiac cycle. The three conditions are: a patent ductus arteriosus (PDA), a leaking sinus of valsalva aneurysm, and a coronary fistula. The patent ductus arteriosus is another fetal bypass of the lungs. The blood goes out the pulmonary artery and travels through a short connection (the (PDA) into the aorta. Within hours of birth, the tube connection closes down under the influence of prostaglandins. Some infants, especially those prematurely born, the tube does not close and remains open. A shunt is then present between the lungs and the aorta. The left atrium is the collecting chamber for the shunt and enlarges in size. This condition can be missed in childhood and later cause trouble from infectious endocarditis or cause elevated pressure in the lungs. A sinus of valsalva aneurysm is an enlargement of the aortic sinus. The aortic sinus has a small buldge that directs blood during diastole into the coronary arteries. There are three in your heart that coincide with the three leaflets of the aorta. The aneurysm can enlarge and weaken with a small hole. The most common location is from the right sinus of valsalva into the right atrium. The right atrium and right heart is the collecting chamber for this condition. Coronary fistulas are abnormal connections of the coronary arteries to pulmonary artery, or other veins. The flow through the coronary is increased and the coronary can get quite large.

Heart Sounds

- S_1 **loud or soft**
- S_2 **split or single, fixed or varies with respiration**
- S_3 **present or not, age of the patient, opening snap, Knock**
- S_4 **present or not - always abnormal ischemia, LVH**

S_1 is the sound the heart makes when the tricuspid and mitral valves close. S_1 can be split into two components mitral and tricuspid. The splitting can be confused with an S_4. The loudness of S_1 is determined by a number of factors.

S_1 LOUD
- **Short PR   Wolf Parkinson White**
- **Pliable Mitral Stenosis OK for Valvuloplasty**
- **Hyperkinetic States**
- **Left Atrial Myxoma**
- **Mitral Valve Prolapse**

S1 <u>soft</u>

- •Long PR first degree AV block
- •LBBB
- •ACUTE aortic insufficiency
- •Flail leaflet
- •Bad ventricle

Electrical coupling of the heart to contraction is a function of the conducting system. The PR interval is the time delay that occurs when an electrical signal passes thru the AV node. The purpose of the AV node is to slow the signal so that the top collecting chamber has a chance to empty into the bottom pumping chamber before it pumps blood out to the body. With a short PR interval the mitral valve is wide open and will shut with vigor causing a loud sound. When the PR interval is long the mitral valve is already drifting close so the door is not slammed shut. The sound is soft. By attention to S1 the electrocardiogram can be predicted. A patient who arrived in the emergency room with complaints of rapid heart beats 10 minutes prior can be diagnosed with WPW (Wolf Parkinson White Syndrome) by hearing a loud first heart sound.

The patient who passed out and regained consciousness one hour prior can give the examiner a clue for the pass out by noting a very soft first sound. Left bundle branch block is a condition when one of the electrical circuits to ventricle from the AV node is damaged. The electrical activity to that part of the heart has to be supplied from the right side by a very circuitous path through non specialized transmission of electricity. The time for contraction is delayed and the heart runs rough. First degree AV block and Left Bundle Branch Block (LBBB) together implies a sick conduction system that could fail and result in a pass out. The second heart sound will be paradoxically split in LBBB. The physical exam can predict the electrocardiogram. The electrocardiogram can in turn predict what the physical exam should be.

Mechanical factors also influence the intensity of the first heart sound. A hyperdynamic ventricle seen in anemia, thyrotoxicosis, post exercise, or hyperdynamic state due to endogenous or drug catecholamines will make S1 loud. The first heart sound will be loud in Mitral stenosis when the valve is pliable. It will be soft if the valve is calcified and has very little movement. Hearing a soft S1 in mitral stenosis means the valve will likely need to be replaced and can not be repaired by surgery or balloon valvuloplasty. A flail leaflet in mitral valve prolaspe due to ruptured chord or a papillary muscle rupture will have a soft S1 because the valve no longer can close. S1 is also loud in the rare condition of atrial myxoma when a mass on the valve causes it to slam shut. Mitral valve prolapse, the most common reason for valve replacement can be detected by a loud first sound, clicks, and characteristic murmur that gets longer with standing. In mitral valve prolapse the collagen is abnomal and the valve becomes redundant too large for the heart. Things that make the heart smaller will exaggerate the mismatch and make the murmur longer. Standing will make the heart smaller and the valve will prolapse into the left atrium to a greater extent allowing more blood to leak back into the left atrium.

S_1 under special circumstances can be variable in its intensity. Careful and prolonged attention needs to be paid to the first sound. All other sounds need to be blocked from attention with the only focus is on the first heart sound.

S_1 variable
- **AV dissociation**
    - **Ventricular Tachycardia**
    - **Complete Heart Block**

In ventricular tachycardia the ventricle speeds along a 150 beats per minute and is doing its own thing. The sinus node and atria is still trying to take control and is running at a normal rate around 70 beats per minute. AV dissociation means the ventricle and the atria are each doing their own thing. The AV interval will change from very short to very long so the first heart sound will vary in intensity. Imagine rapid fire machine gun fire that gets closer and farther away like a drunk marksman.

Another form of AV dissociation occurs when the heart is going very slow due to complete heart block. The ventricle will be going 20 to 30 beats per minute and the atria is going normal at 70 beats per minute. Imagine a slow lub dub with lub changing in intensity.

S_2 single or split (splitting is wide, fixed or varies with respiration)

S_2 is the second heart sound and is generated by the closure of the aortic and pulmonary valve. The aortic valve normally closes first because the systemic pressure, the afterload resistance is high and will slam the aortic valve closed. The pulmonary valve will close after the aortic valve because the pressure in the lungs and afterload resistance is very low. The timing of the aortic and pulmonary closure can be variable and is also determined by electrical activation of the left and right heart and by mechanical means.

- **Increased splitting with inspiration**
- **Fixed split   atrial septal defect**
- **Widely split RBBB (deep Expiration)**
- **Paradoxical with LBBB**
- **A sign of cardiac dys-synchrony**

Respirations will mechanically alter the closing of the second heart sound. As we inhale, the pressure in the thorax decreases. The lower pressure sucks more blood into the right heart. Since there is more blood in the right heart, it takes longer for the pulmonary valve to close. During exhalation there is less blood in the right heart and the pulmonary valve closes earlier at about the same time as aortic valve closure. This is known as increased splitting of the second heart sound with inspiration. In right bundle branch block the contraction of the right ventricle is delayed so the pulmonary valve closes even later than normal. The second heart sound is now widely split. This can be confused with a fixed split sound.

The importance of distinguishing a wide split versus a fixed split can be measured in dollars. You are examining a patient who has a widely split second heart sound. The patient may have an Atrial Septal Defect that is characterized by this finding. In an atrial septal defect there is a hole in the upper chambers between the right and left atrium. Since blood can pass easily into the right and left atrium there is no preferential flow into the right heart when the pressure falls in the chest during inspiration. The second heart sound is fixed split and only depends on the differential in pulmonary and aortic afterload. If you send this patient to get an echocardiogram it will cost thousands of dollars. The alternative is to simply have the patient exaggerate their breathing by taking a deeper breath in, and more importantly, a deeper breath out. In right bundle branch block, a deeper breath out will close the splitting of the second heart sound confirming there is no atrial septal defect.

Pulmonary hypertension and aortic stenosis will both mechanically alter the timing of the aortic and pulmonary closure. The resistance to flow will cause a time delay for closure as ventricle struggles to get the blood through the obstruction. In aortic stenosis if the second heart sound is single the aortic stenosis is considered severe. The second heart sound can be single in the condition of aortic stenosis for two reasons. The first is that the valve is so stiff that it never opens and never closes and thus does not make a sound. The more common reason is the left heart struggles mechanically to expel the blood and contraction is delayed allowing the aortic and pulmonary valve to close at the same time. Either of these indicated a severe stenosis. Other reasons for a single second heart sound include conditions such as absence of the pulmonary valve. This condition exists in surgically corrected Tetralogy of Fallot when the pulmonary valve is excised to relieve infundibular stenosis. If the valve is not present it will not make a sound.

With pulmonary stenosis, the timing is not as important as the loudness of the P_2. The pulmonic closure sound is considered loud if it can be heard at the apex. The loudest second heart sound auscultated over the pulmonic area is a condition of two wrongs nearly making a right. Congenitally Corrected Transposition is a congenital heart defect that may allow the individual to live a perfectly normal life, sometimes without the individual even knowing they have the condition. On the other hand, transposition of the great vessel is a near fatal condition better known as blue babies. Transposition is the switch of the pulmonary artery and the aorta. There are now two parallel circulations with the right heart pumping blood to the aorta into the body with return of the blood to the right heart. The left ventricle pumps blood to the lungs and the blood returns to the left ventricle. What allows the babies to survive is persistence of the fetal circulation, and atrial or ventricular septal defects that allows mixing of the two parallel circulation.

Congenitally Corrected Transposition has two errors that nearly cancel each other. In addition to the transposition of the great vessels the ventricles are inverted so that the right ventricle pumps blood to the aorta and then the blood is returned to the right atrium that is now attached to the left ventricle that pumps the blood to the lungs. These two wrongs allow for serial instead of parallel circulation. Both conditions have the aortic valve in the pulmonic position anterior in the chest cavity. The systemic valve is just under the chest wall and the auscultation will reveal an extremely loud second heart sound. These patients may be perfectly fine for their entire life while others will develop heart failure since the right ventricle was not design to pump into a high pressure system.

The Houckster was awakened at 2 A.M. with a phone call. The night had already been bad since he had experienced his first earthquake. He was dreaming he was at home with is wife and the wave motion of the bed seemed natural. The silverware dancing across the tabletop was the first clue that not all was right. The earthquake at Clark Air Force Base was an early sign of the eruption that would occur. The earthquake was at 11PM. After collecting his rattled senses he recognized the frazzled voice of the neonatologist. On arrival, the neonatoligist was describing a baby whose saturation could not be maintained greater than 60% despite 100% oxygenation. The baby had no murmurs but the second heart sound was very loud. The echocardiogram revealed after some difficulty a shotgun appearance of the great vessels indicating transposition. This was confirmed when careful imaging of the aorta revealed a bifurcation confirming it was not the aorta but the pulmonary artery. The deadliness of the condition was punctuated by the demonstration of the atrial and ventricular septums were intact. The only mixing of blood was through the patent ductus arteriosus that could close at any time. Medications were being given to keep this small lifeline open. Tripler Army Hospital was alerted but they really did not believe the adult cardiologist and the neonatolgist who were 8,000 miles away. It would cost $55,000 to Medivac the new born to California. On arrival to Hawaii the baby was assessed and immediately sent to California for surgery. A loud pulmonic closure sound suggests transposition.

The timing of aortic and pulmonic closure depends on mechanical and electrical influences. As we learned above, right bundle branch will delay the activation of the right ventricle and thus will delay the closure of the pulmonary valve. This can be confused with an atrial septal defect. Left bundle branch block will cause a similar delay in activation of the left ventricle. This causes a soft first sound and delayed closure of the aortic valve. The aortic valve will now close after the pulmonic valve. Inspiration will again delay the closure of the pulmonary valve. Instead of inspiratory splitting there is paradoxic inspiratory closure of the second heart sound. This finding is also the physical exam finding for cardiac dys-synchrony.

S_3 present or not

age of the patient
distinguish from opening snap or a pericardial knock

An S_3 is the sound that is produced when blood rapidly fills the ventricle during diastole. This can be a normal finding in a youth or an athlete who has a dynamic ventricle that can suck blood into the ventricle. In older patients it is a sign that the preload is too high forcing blood into a full ventricle, a sign of congestive heart failure. An S_3 in an adult always means a diuretic is needed. Athletes and children eventually lose the S_3 as they age.

Sounds that occur at nearly the same time as an **S₃** include a pericardial knock due the stiffness of the ventricle from constrictive pericarditits. Rheumatic mitral stenosis has an opening snap that also occurs in diastole. The "knock" and "snap" sound much different from an **S₃** and both have higher frequency components. A third heart sound can be brought out by having the patient make two fists to increase afterload. The increased stiffness of the circulation stiffens the ventricle and brings out the filling sound. The bell amplifies an **S₃** and can be extinguished by pressing the bell harder against the chest wall.

S₄ present or not
> **always abnormal**
> **ischemia, LVH, restrictive heart disease**

The fourth heart sound occurs late in diastole. The sound is generated when the atrium forces blood into a non-compliant left ventricle. To generate a fourth heart sound the patient has to be in sinus rhythm. The sound always indicates a disease is present. Hypertension with pathologic hypertrophy (thickening) of the heart wall muscle is a common cause. Ischemia, myocardial infarction will cause a build up of calcium within the cells. The binding of actin and myosin will not release with the extra calcium. The fibers no longer slide but are ratcheted together like Velcro.

Extra cardiac sounds

•**Ejection Clicks** (bicuspid aortic valve, pulmonary stenosis)

A bicuspid aortic valve is the most common congenital anonomaly. Two of the three raphe are fused so when the valve opens it looks like a fishmouth instead of a triangle. The ejection click is a clue this is present. Later in life the valve may become stenotic or may begin to leak.

The pulmonary valve can also become stenotic and will click. This click is loudest during expiration and is the only sound that gets louder during expiration.

•**Clicks** (mitral valve prolapse)
Mitral valve prolapse is a larger than necessary valve so a portion of the valve prolapses into the atrium. The resultant clicks occur when the chordae are tensed by the bull whip action of the redundant leaflets. The clicks move with maneuvers. When the patient squatts the clicks move toward the second heart sound. With standing they move into the first heart sound and may disappear. The murmur usually gets longer with standing, but is not always present. The murmur depends on the volume status of the ventricle.

•**Honks, whoops** (mitral valve prolapse) - Other interesting musical noises

•**Opening Snaps -** rheumatic mitral stenosis

Mitral stenosis was once very common and there were wards of patients suffering from this condition. Strep throat would initiate an immune reaction that would cause continued damage and fibrosis to the mitral valve. The valve would get thicker and the opening would become reduced. The murmur is very soft and difficult to detect, but the opening snap, the tug on the thickened leaflets is easy to hear. The opening snap can be heard in the seated position and if it is heard the patient can be placed into the left lateral decubitus position. In a quiet room the diastolic blowing soft murmur can be heard if you ignore the loud mitral regurgitation murmur.

•**Rubs** (pericarditis)

A pericardial friction rub is the sound a saddle makes as you are riding a horse. The inflamed pericardium looses its lubricant action that allows the heart to move in the chest. The patient can feel this loss of lubrication and suffers a rhythmic chest pain.

It is now time to get back to the original Sherlock Holmes teaching point. By using palpation of the carotid artery to determine the state of the systemic circulation, palpation of the systolic pulsation of the pulmonary artery to determine the state of the pulmonary circulation, and palpation of the point of maximum impulse to determine the state of the ventricle, all valvular abnormalities can be detected without auscultation. The purpose of auscultation is to confirm your diagnosis and to determine the severity of the valvular abnormality. A test of what you learned earlier in the chapter will now begin. Fasten your thinking cap and put your brain in gear. There will be five common conditions presented and your job is to determine what valvular abnormality is responsible. Do all five prior to looking on the next page which will give the answer. You are encouraged to look back at pages 140, 141 that you had ear marked previously to help in your decision process. Remember to put brain in gear and then analyze the following:

State of the Systemic Circulation	**Carotid Pulsation**
State of the Pulmonary Circulation	**PulmonaryTap**
State of the Ventricle	**PMI**

Condition 1   ?

State of the Systemic Circulation	Decreased Carotid Pulsation
State of the Pulmonary Circulation	Normal Pulmonary Tap (–)
State of the Ventricle	PMI displaced discrete

Condition 2   ?

State of the Systemic Circulation	Decreased Carotid Pulsation
State of the Pulmonary Circulation	Pulmonary Tap (+) Positive
State of the Ventricle	PMI Small

Condition 3   ?

State of the Systemic Circulation	Increased Carotid Pulsation
State of the Pulmonary Circulation	Pulmonary Tap (–) Negative
State of the Ventricle	PMI displaced diffuse/discrete

Condition 4   ?

State of the Systemic Circulation	Increased Carotid Pulsation
State of the Pulmonary Circulation	Pulmonary Tap (+) Positive
State of the Ventricle	PMI displaced diffuse/discrete

Condition 5   ?

State of the Systemic Circulation	Decreased Carotid Pulsation
State of the Pulmonary Circulation	Pulmonary Tap (+) Positive
State of the Ventricle	Displaced Diffuse

Condition 1	Aortic Stenosis
State of the Systemic Circulation	**Decreased Carotid Pulsation**
State of the Pulmonary Circulation	**Normal Pulmonary Tap (–)**
State of the Ventricle	**PMI displaced discrete**

The diagnosis of Aortic stenosis is suggested by a delay in carotid upstroke with low volume. If a thrill is present the gradient is likely greater than 50 mmHg. The lack of a pulmonary tap means the mitral valve is functioning and the lungs are protected. In late aortic stenosis the ventricle gets larger and it is natural for the mitral valve to leak and the pulmonary pressures may become elevated late in the course of aortic stenosis especially if heart failure symptoms are present. The only valvular abnormalities that gives a delayed upstroke are aortic, mitral, and pulmonary stenosis. Mitral stenosis, pulmonary stenosis and a bad ventricle all have associated pulmonary tap which is absent in condition 1. The PMI demonstrating a hypertrophied ventricle gives us further confidence that aortic stenosis is the diagnosis.

What should the examiner listen to in order to determine the severity of the condition? The murmur will be loud, systolic, and if it is late peaking the ventricle is struggling to push the blood out of the valve. A single second heart sound confirms the severity.
The physical exam should be described as follows: the carotid upstroke was delayed and low volume. There was a thrill present. Pulmonary systolic pulsation was absent and the PMI was mildly displaced and discrete consistent with left ventricular hypertrophy. My estimate of the ejection fraction is 60%. A late peaking systolic murmur was present with a single heart sound consistent with critical aortic stenosis. The first heart sound was soft and there was no third or fourth sound.

Condition 2	Mitral Stenosis/ Pulmonary Hypertension
State of the Systemic Circulation	**Decreased Carotid Pulsation**
State of the Pulmonary Circulation	**Pulmonary Tap (+) Positive**
State of the Ventricle	**PMI small**

There are two possible answers. Mitral stenosis would be your first guess and pulmonary hypertension would be your second guess. The decreased carotid again limits your choices to a bad ventricle, mitral stenosis, and pulmonary hypertension due to primary pulmonary hypertension or due to pulmonary stenosis. All of these conditions also have a pulmonary tap associated. The left ventricle being small then limits your choices to mitral stenosis or pulmonary hypertension.

What should the examiner listen to in order to determine how significant the mitral stenosis? A loud first heart sound will indicate a pliable leaflet that could be repaired or referred for balloon valvuloplasty. The severity of mitral stenosis is determined by the P_2 opening snap interval. The shorter timing of the interval the worse is the stenosis.

This finding is no longer reliable if a prior mitral valve intervention had been performed. Auscultation of no murmurs but an accentuated pulmonary closure will confirm pulmonary hypertension

The physical exam would be described as follows: The carotid upstroke was diminished with low volume. There was no delay in the upstroke. The pulmonary artery had a prominent systolic pulsation consistent with pulmonary pressures greater than 50 mmHg. The PMI was small and non-displaced. The first heart sound was loud and consistent with a pliable mitral valve. The second heart sound had a prominent P_2 followed by an opening snap that followed closely after P_2 indicating severe mitral stenosis. A 3 out of 6 holosystolic murmur was heard and in the left lateral decubitus a short I/VI diastolic murmur with a pre-systolic rumble was heard. Not always but frequently mitral stenosis also leaks.

Condition 3	Aortic Insufficiency
State of the Systemic Circulation	Increased Carotid Pulsation
State of the Pulmonary Circulation	Pulmonary Tap (–) Negative
State of the Ventricle	PMI displaced diffuse/discrete

The physical exam finding of dynamic carotid upstroke could be due to systemic reasons for increased cardiac output such as anemia. The valvular abnormalities of mitral regurgitation and aortic insufficiency are also possible. If the pulmonary tap can not be felt, mitral regurgitation can be eliminated leaving only aortic insufficiency as the correct answer.

In this case, the examiner will hear a systolic murmur because of the increased flow across the aortic valve. If 2 liters are falling back then 7 liters have to be forward. This will cause turbulence across the valve. The softer diastolic murmur will not be easy to hear and will require concentration and ignoring the louder systolic murmur. The length of the diastolic murmur determines the significance of the aortic insufficiency. A holo-diastolic murmur is severe aortic insufficiency. These findings are for chronic aortic insufficiency. If the aortic insufficiency is acute due to destruction of valve from endocarditis there may not be a murmur. In acute aortic insufficiency, the aorta's diastolic pressure falls as the left ventricular pressure rises. The gradient decreases so the murmur may be non-existent despite a torrential leak.

The physical exam would be described as the carotids hyperdynamic in upstroke with full volume. There was no systolic pulsation of the pulmonary artery, and the PMI was displaced lateral but was still discrete. The ejection fraction was normal. Examination of the nailbed demonstrated a flashing pink Quincke's pulse. A slight head bob was present. These last two findings are peripheral findings due to the large volume of forward blood with rapid fall off. The large volume of blood can make the head bob, a crossed leg bob, and cause pistol shot femoral pulses. It was hypothesized that Abraham Lincoln had Marfan's syndrome and aortic insufficiency since every one of Brady's photographs always had a blurred bobbing crossed leg.

Auscultation reveals a soft first sound, an ejection murmur grade III/VI in intensity and a long diastolic murmur that stretched two thirds of diastole. There was no third or fourth heart sound.

Condition 4	Mitral Insufficiency
State of the Systemic Circulation	Increased Carotid Pulsation
State of the Pulmonary Circulation	Pulmonary Tap (+) Positive
State of the Ventricle	PMI displaced diffuse/discrete

Increased carotid pulsation with a pulmonary tap suggests mitral regurgitation. Mitral regurgitation is associated with mitral valve prolapse and accompanies inferior myocardial infarctions. Any ventricle that becomes dilated from heart failure will have a stretch of the annulus with resultant mitral regurgitation. The murmur is always loud and the only way to judge significance is to detect the pulmonary tap.

The physical exam would be described as the carotid had a dynamic upstroke. Systolic pulsation of the pulmonary artery was present indicating pulmonary hypertension. The ventricle was displaced with a diffuse PMI suggesting a less than normal ejection fraction estimated at 45%. Auscultation of the chest revealed a soft first heart sound. With hand grip a soft intermittent third heart sound indicating volume overload and failing ventricle.

Condition 5	CHF
State of the Systemic Circulation	Decreased Carotid Pulsation
State of the Pulmonary Circulation	Pulmonary Tap (+) Positive
State of the Ventricle	Displaced diffuse

The most common form of heart disease today is congestive heart failure. History should suggest the diagnosis with symptoms of shortness of breath when one lies down at night, shortness of breath with exertion, and leg swelling. The diagnosis is made by noting that the forward cardiac output is poor with decreased carotid upstroke. The filling pressures of the heart are known to be increased when systolic pulsation of the pulmonary artery is palpated. The diagnosis is confirmed when the ventricle is found to demonstrate poor systolic function with a displaced and diffuse ejection fraction.

The severity of the left ventricular dysfunction can be estimated by the pulmonary tap and by the estimate of the ejection fraction. While in most conditions, the ejection fraction determines the patient's prognosis, with very poor ejection fractions, this number is no longer predictive. The filling volumes of the heart become a better predictive factor. The ultimate predictor is not a physical exam finding but a measure of renal function.

Alfredo was an 80 year-old male with several medical conditions. The patient had congestive heart failure for over 15 years. He had been placed on medical therapy and had near improvement in his cardiac function. He had abdominal aortic aneurysm and a thoracic aneurysm. As a result of his thoracic aneurysm he had a moderate amount of aortic insufficiency. The left ventricular dysfunction was thought to be due to long-standing aortic insufficiency. Initially he presented with an ejection fraction of 20% and surgical correction was not considered because the ventricle was too far gone. With medical therapy his ventricle recovered to an ejection fraction of 50% and surgery at this time was not considered because he had no symptoms and the aortic insufficiency no longer seemed to be the culprit. Conditions were improving until Alfredo experienced an episode of sudden cardiac death from which he was resuscitated. The defibrillator was placed which caused cardiac dyssynchrony and depressed the ejection fraction to 25%. This device was upgraded to a biventricular pacemaker defibrillator.

Over the next two years, Alfredo's blood pressure continued to fall and he no longer was on as much medication. Upon his visits to the clinic, it was not common for his blood pressure to be 80 systolic. Although he rarely suffered from congestion, Alfredo suffered symptoms of low output with fatigue, weakness, and dizziness with standing. He was admitted to the hospital with a systolic blood pressure of 60. Further attempts at adjusting his medications to relieve this condition failed. The patient and family decided the best option was to be placed on hospice. He was discharged from hospital after his defibrillator had been turned off. Six months later he returned to clinic and was discharged from hospice because he would not die. His blood pressure was 120 systolic. The defibrillator was again turned on and his medications were titrated. This patient continued to do well and the reason for his previous decline remains unknown. A theory of endocrine failure or vasodilatation due to toxins from edematous bowel may have caused his low blood pressure. Perhaps low body temperature would not allow pro-molecules to be cleaved into active forms. Predicting his mortality was fairly easy because he had all of the requirements for mortality including a depressed ejection fraction a dilated left ventricle and poor renal function. Even with these strong predictors determining the patient's final demise was not possible.

The physical examination described the skin as cool to the touch and exhibited a ghostly pallor. The carotids were diminished in upstroke indicative of poor cardiac output. Systolic pulsation of the pulmonary artery was present with an estimate of right heart pressures exceeding 50 mm mercury. The point of maximum impulse was displaced into the lateral chest and was large and diffuse. The first heart sound was soft and the second heart sound was paradoxically split, consistent with left bundle branch block. A holosystolic murmur of mitral regurgitation was present in a mid-axillary line and radiated throughout the chest. A gallop rhythm was present consisting of summation of the third and fourth heart sounds.

The above Sherlock Holmes method of physical examination can be applied to every patient. The disease process, however, extends from normal to very abnormal. The physical exam findings may be very subtle in patients who only have minor problems as opposed to those patients who are at the end stage of their disease. The learning process is one of evaluating many patients and making many mistakes and learning from your mistakes. With each patient evaluation, physical examination skills will become better.

From time to time, everyone becomes distracted resulting in poor physical examination skills. It is important to document your findings so that you can compare your examination to prior examinations. Patients can change and consequently their examination will change.

One last caution is that all findings should be consistent. Take the example of a delayed carotid upstroke in the presence of the pulmonary tap and a hypertrophied ventricle. There is no single diagnosis that can explain all of these findings. The most likely diagnosis is aortic stenosis. The presence of a pulmonary tap would suggest that there is heart failure or mitral valve disease present. The natural history of aortic stenosis is left ventricular failure and resultant mitral regurgitation. The above findings are therefore consistent with aortic stenosis and mitral regurgitation. Both of these findings will result in a loud systolic murmur. The examiner must determine whether both are present. A maneuver is performed with handgrip where the patient makes to fists and squeezes as hard as possible while he continues to breathe normally. The handgrip will increase systemic vascular resistance and decrease the flow across the aortic valve with the resultant reduction of the aortic stenosis murmur. This same maneuver will increase the amount of mitral regurgitation and increase the intensity of this regurgitant murmur. By performing this maneuver and listening in the axilla and over the aortic area both of these conditions can be confirmed.

There are many other physical examination pearls that can be learned from textbooks. Auscultation can be learned through the Blaufass.com or ACC.com heart tones web site. The real art of physical exam is to constantly test your skills. The physical examination will improve over time and after exposure to diseased states. Some of the rare findings may take years to surface into your practice. The skill of physical exam will improve with the number of examinations that are performed. I would like to end this chapter with an important final pearl.

When doing school athletic physicals, the student should always be listened to in the standing position. Deaths during athletic activities are quite rare and should not be feared. A common condition in those athletes who do die is hypertrophic cardiomyopathy with obstruction. In this condition, the septum is hypertrophied and causes a narrowing of the left ventricular outflow tract. The anterior leaflet of the mitral valve acts like an airplane wing and flies across the outflow tract and obstructs the flow. This condition is aggravated when the heart is smaller. The murmur gets much louder with standing because the heart gets smaller and may not be present if the patient in the supine position. It is not desirable to have your name on the physical examination form of a child that died from sudden cardiac death during athletic activity. Even some of these children may have negative exams. If you at least documented that you attempted to find this condition by listening while standing you have demonstrated every attempt to be a good physician.

The next chapter is complimentary to the physical exam and will explore hemodynamics. Hemodynamics is the cause of physical exam findings.

Chapter 11
Hemodynamics – Cardiac Performance Parameters How It All Works

Understanding how it all works requires a good understanding of hemodynamics, the study of blood flow throughout the heart and body. The study of hemodynamics requires the cardiac performance parameters are known as well as an understanding of the interaction between the pump and the blood vessels. The greatest efficiency of the pump occurs when the cardiac performance parameters are optimized. When I first started giving this lecture I listed only four performance parameters. While writing this chapter, I raised the number of parameters to six with the last one broken into two parts. Before finishing the chapter I increased the parameters to 7. When I ask medical students to define the parameters, they always volunteer heart rate as one of the cardiac parameters. Although it is well known heart rate is important in blood flow, it plays an extrinsic role in cardiac performance. This extrinsic role is not considered one of the cardiac parameters. Heart rate is important because if the heart goes too slow the amount of flow to the body is diminished. In turn, if the heart rate goes too fast there is insufficient filling of the heart so flow to the body is again diminished. The cardiac parameters are listed as follows:

1)   **Preload**- Preload is the stretch of the myocardium and is determined by measuring the volume of the heart or indirectly by measuring the pressure in the heart chamber.

2)   **Afterload**- Afterload is the size of the arterioles. It is the resistance the heart encounters when it is pumping into a vascular space, the resistance to flow. The afterload determines the amount of work the heart has to perform. It is more difficult to pump into miniscule arterioles then into large dilated arterioles.

3)   **Contractility**- Contractility is a systolic function and is the squeeze of the cardiac muscle. The fancy word for contractility is inotropy

4)   **Compliance**- Compliance refers to the stiffness of the ventricle and is a diastolic function. A very thick walled balloon is more difficult to blow up than a thin walled balloon. Compliance is also known as lusitropy.

5)   **Neuroendocrine**-The heart does not exist by itself, but in the context of the body. The neuroendocrine system regulates both the heart function and the body's response according to certain stressors. The neuroendocrine system allowed us to crawl out of the ocean. Some people believe it was the lungs that prompted the exit from the sea to the land. The evolution of the kidney allowed primitive organisms to move to the land. The kidney assured salt and water conservation. The endocrine system also regulated the size of the blood vessels and contributed to salt, water, and volume preservation. This system was the lifeline to protect the body from injury during hemorrhage. The kidney is the most primitive organ. The kidney is stupid because it is so old!

6)   **Geometry and Synchrony**-Geometry is the shape of the heart. If we examine the giraffe's heart, the heart is elongated so it can pump blood 25 feet into the air. A reptilian heart in contrast is spherical. Reptiles remain flat on the ground because they cannot generate blood pressure. A heart failure patient will develop a spherical heart as the heart becomes weaker. Some heart failure patients should not be more than 4 inches off the ground. Synchrony referrers to the electromechanical coupling of the heart. The electrical conduction system tells when the heart is to contract and in what sequence.

Damage to the conducting system is similar to a V-8 engine with several of spark plugs removed. The heart will run inefficiently in this unsynchronized fashion.

7)   Blood Vessel Properties - This is the last parameter and in the past was labeled by individuals of my parents' age as hardening of the arteries. This term better defines the properties of blood vessels than atherosclerosis. Atherosclerosis is hard to say and makes you think of cheese. Hardening of the arteries refers to the stiffness of the vessel. Just as sound waves travel through the air, pressure waves travel through the blood vessels along with the flow of blood. These waves reflect from branch points and can add pressure to the blood vessel. The heart then sees a higher pressure just because the blood vessel is stiffer. This is the likely cause of 50% of diastolic heart failure.

Each of these parameters will be examined in more detail, emphasizing the physical examination of the parameter and how one can change the parameter with medications or maneuvers. By understanding the parameters and how they can be changed, every kind of heart disease can be approached and treated. Later in this chapter a number of different challenging cases will be presented to emphasize some specific points in the care of patients.

Preload

How many preloads does your heart have? If you are thinking that you only have one heart you are incorrect. We have two hearts: one is the right heart, which pumps blood to the pulmonary circulation to re-oxygenate our blood. There is a second heart, the left heart that pumps blood to the body supplying it with nutrients and oxygen. Placing into a flask the separate contractile protein elements of the heart muscle, Actin and Myosin (defined as the thin and thick filaments); the Actin and will bind to the Myosin forming a solute. To get them to unbind, Troponin is added to the flask and this protein will cause the Actin and Myosin to go back into solution. If we then wanted the Actin and Myosin to bind we would add Calcium to the flask that would inhibit the inhibitor Troponin allowing the Actin and Myosin to bind. No active contraction so far has occurred in this process. For the contraction to occur the Myosin must change configuration through an energy utilizing step. In this process ATP is converted to ADP plus Phosphorous. The strength of contraction is dependent on the number of binding sites of the Actin and Myosin elements. If the muscle is stretched too far there will be binding sites where there is Actin but no Myosin. If the muscle is not stretched enough there will be overlap of Myosin with insufficient Actin. An experiment can be performed with a cat papillary muscle that demonstrates the perfect length stretch to be 2.2 μ for the strongest contraction where all Actin and Myosin are completely bound.

Starling's Law of the Heart refers to an entire intact heart where the best performance occurs at a certain volume. The volume of the heart can be different between individuals depending on size of fibers and the amount of interstitial fibrosis. Starling's law states there is an optimum preload volume where performance of the heart is greatest. This determines the stretch of the heart.

Although the volume is the preferred measure in Starling's Law of the Heart, it can not be measured easily in the coronary care unit. Instead pressure is measured and the volume is inferred by a linear relationship. The height of a column of fluid is a measure of pressure. The height of a dam determines the stored energy or volume of water it holds. The physical exam finding for pressure of the right heart is the height of the Jugular veins. This is a direct measure of preload since pressure is directly related to volume. Visualizing the distention of the jugular veins in the neck is a direct measure of the pressure of the right heart. By estimating the height of this column of fluid the volume that stretches the myocardium is inferred.

A direct measure of left heart preload by visualizing a column of fluid is not possible. There are two physical exam findings for left heart preload. The first is a nonspecific finding and is known as rales, crackles or rhonchi. These are the noises that the air spaces make when they begin to fill with fluid. This is a nonspecific finding because in some individuals rales may be present due to interstitial lung disease. In chronic heart failure, the lungs may remain clear even though the filling pressures are quite elevated. Heart failure patients adapt to a high filling pressures to keep from drowning. As the pulmonary venous pressure rises, Starling's forces imply that there would be more interstitial edema fluid moving into the lungs. The heart failure patient is able to adapt by increasing the size of the lymphatic channels that will rapidly drain the fluid as it accumulates. These lymphatic channels can be seen on chest x-ray as Kerley B lines.

Preload can be increased or decreased. Increasing preload is appropriate in patients who have become dehydrated because of hemorrhage, diarrhea, vomiting, lack of intake of nutrients, or shock. The absolutely best fluid to increase preload is blood. Blood is cellular and remains in the vascular space for extended periods of time. All other fluids will only have transient effects on the volume of the vasculature. The next best fluid is colloid consisting of albumin, a blood byproduct. Albumin is a protein which remains in the blood vessel longer slowly metabolizing than the third choice which is crystalloid. Crystalloid, or saline, is essentially seawater getting its name from its salt content. There are two types of saline that can be used. Ringer's lactate is a favorite among surgeons and medicine doctors tend to use sodium chloride. There is no difference in how these solutions change preload. There is a difference depending on volume of solution required. Patients who receive large dosage of sodium chloride (more than 3 to 4 liters) will develop acidosis. Ringer's lactate is preferred because the lactate will be converted to bicarbonate preventing this metabolic acidosis. For small volume resuscitation there is no difference.

Lowering the preload quickly is necessary when there is severe shortness of breath associated with congestive heart failure. High preload makes the lungs heavy and makes you feel like you are drowning. Lowering preload quickly in this dire circumstance is very desirable. Gravity can be used to lower preload by sitting or standing up. This is why EMS carries the heart patients in a sitting position. Dialysis can lower preload by removing fluid. Phlebotomy, or bloodletting, is an age old trick to quickly lower preload. If a dialysis machine is not available, a unit of blood can be drawn off the patient and taken to the blood bank, spun down with the packed cells returned to the patient. Rotating tourniquets is no longer practiced. The machine designed to perform this task was quite cumbersome. A blood pressure cuff was placed on each extremity and inflated in order to decrease the return of blood to the heart from the nonessential limb. The nonessential limb would become ischemic if the cuff was inflated too long so the cuff on that limb would be released while the next limb's cuff would inflate.

For functioning kidneys, diuretics, like Lasix, can remove fluid at a slower rate. A continuous infusion of Lasix is the most efficient means between 5 and 20 mg per hour. Nitroglycerine is a venodilator as well as morphine. Natrecor will lower the preload faster than nitroglycerine or Lasix and will more rapidly improve the drowning sensation of heart failure.

Afterload

Afterload is the resistance to flow, the size of the arterioles. There are two afterloads in your heart. The left heart pumps into the systemic circulation, the systemic arterioles. The physical examination of this parameter is the carotid upstroke. A brisk full upstroke suggests a good cardiac output and low resistance. A diminished carotid means poor cardiac output and high afterload. Other helpful signs are skin warmth, pallor, and urine output most of which reflects cardiac output. The right heart afterload consists of the pulmonary arterioles. These arterioles branch many times with ever increasing volume. This branching and increased surface area makes the right heart afterload very low. The physical exam finding of right heart afterload is systolic pulsation of the pulmonary artery. If palpated this finding is always abnormal. If a pulmonary tap can be felt the pulmonary pressures are likely high in excess of 50 mmHg. Review the physical exam chapter on Systolic pulsation of the pulmonary artery. Afterload is reduced by increasing the size of the arterioles.

Two patients – One patient (A) has a blood pressure of 120/80 and patient (B) has a blood pressure of 220/80. Which one has the highest afterload? The question is a trick since from the information provided there is no answer. A new student would think it was patient (B) because the pressure was very high. The degree of blood pressure does not tell you the size of the arterioles. Patient (A) was suffering from congestive heart failure. The ejection fraction was 25%. His skin was cool and poorly perfused. The carotid upstroke was markedly diminished and poor in volume. His cardiac output was three liters per minute. The arterioles are markedly constricted attempting to maintain blood pressure with a limited cardiac output. Patient (B) was a 17-year-old running at the 15th minute of Bruce protocol with his arterioles completely dilated to accept as much blood as possible while he was performing to his maximum capability on the exercise treadmill test. His cardiac output was 20 L per minute. A formula for afterload follows:

Afterload   SVR = (Mean BP - CVP) * 80/CO

SVR = systemic vascular resistance
Mean BP = mean blood pressure (2 x diastolic +systolic)/3
CVP = preload usually ignored because it is low
CO = cardiac output

(A) - EF 15% Pulmonary Edema cool skin   SVR = (93.3-10)*80/3 = 2222
(B) - 15 y/o 18 min. of Bruce ETT   SVR = (126.6-5)*80/20 =486

The equation can be rewritten **Cardiac output = Mean BP*80/SVR**
This equation states that cardiac output can be increased by lowering the systemic vascular resistance- even if the blood pressure is low. Do not be afraid of gradually increasing ACE inhibitors.

Afterload is reduced with arteriole vasodilators which include Nitroprusside, Fenoldapam, ACE inhibitors or blockers, Natrecor, Hydralazine, and Nitrate combination. When patients are critically ill the use of short acting agents, such as Nitroprusside, should be used. The agent can be increased quickly and will go away fast if there is an adverse response. Nitroprusside can not be used for long periods of time because of Cyanide toxicity. Cyanide toxicity will occur quicker if there is renal insufficiency. Fenoldapam is a Dopaminergic vasodilator and can be used with patients who have renal problems. Natrecor is my favorite arteriole dilator because it has all of the attributes of a heart failure medications. It lowers preload fast. It lowers afterload. It improves compliance. It blocks all of the neurohormones ACE, Aldactone, and sympathetic hormones and is natruretic eliminating sodium causing a diuresis. Of all of the different ACEI, Captopril should be used because it has the shortest half-life and can be titrated faster. Long acting ACEI or ACE blockers can be used after titration so the patients will be more compliant with once a day medications. Hydralazine and nitrates likely work through the nitric oxide pathway and is especially favorable in hypertensive heart disease and African Americans because they have a predominant hypertensive heart disease.

Mechanical afterload reduction is provided by a balloon pump. This 30 to 40 cc balloon inflates during diastole when the heart is resting and deflates during systole the time blood is ejected from the heart. The heart pumps into 40 cc's of empty space. A vacuum is a very good afterloading agent. The balloon pump will augment diastolic pressure and provide more blood flow to the heart. The intra aortic balloon can not be used when aortic insufficiency is present.

Afterload must be increased when the arterioles vasodilate inappropriately or blood pressure falls. Shock can occur from volume depletion, bacterial infection (sepsis), and from evil humors after bypass surgery or allergic reactions. There are a number of different agents that can increase afterload until the reason for the shock is corrected.

One class of agents is the sympathomimetic amines, a fancy language for adrenal hormones. The adrenal hormones are the fight or flight hormone that can increase blood pressure, heart rate, and contractility. They each have special properties. Epinephrine is the agent of choice in anaphylactic shock due to an allergic reaction. It will raise blood pressure, raise heart rate, and will cause acidosis because of its intense vasoconstriction in the periphery. All organs will be affected. Dobutamine has similar properties but has more vasodilator properties so it is helpful in cardiogenic shock with high afterload. It will increase heart rate to a lesser amount as compared to epinephrine. Dopamine is similar to Dobutamine in increasing contractility but is a vasoconstrictor and is used if blood pressure needs to be raised. Levophed (leave them dead) is similar to dopamine but has even more intense vasoconstriction. Levophed seems to be the agent of choice in septic shock. Neosynephrine is a vasoconstrictor that does not increase heart rate. It is very useful in patients who already have a high heart rate.

The second class of agents, vasopressin, is more favorable to cardiac disorders. The mechanism of action is not well known. Although providing vasopressin to a normal subject will not elevate blood pressure, administering it during shock will raise blood pressure by constricting non-essential organs and preserving flow to the kidney. There are separate receptors that can cause vasodilatation and constriction giving a favorable outcome for survival. It is easy to dose. Give 40 units intravenously if the patient is totally dead. Give 20 units if only half dead. Give 4 units if really sick and start a drip of 1 to 2 units per hour. If you want to preserve organs the kidney and the brain give vasopressin as your first choice of vasoconstrictor in patients with a bad heart pump.

Placing tourniquets around the limbs will squeeze the vessels and produce a mechanical afterload. Squatting will kink the great vessels and increase afterload. It will also increase preload and make the heart bigger. Tetralogy of Fallot patients squat to get more blood flow to the lungs. Hypertrophic cardiomyopathy will have less obstruction during squatting.

Afterload reduction of the pulmonary vasculature can be accomplished by Nitroprusside, Adenosine, Nitric Oxide, Flolan, calcium channel blockers, and Viagra. Specialized doctors in treatment of pulmonary hypertension make highly educated and accurate guesses as to which agent to use.

Contractility (Inotropy)

Contractility is the squeezing function of the heart. It is not be measured with a Swan-Ganz catheter and contractility can only be estimated indirectly by the ejection fraction. Contractility is determined in the lab by pressure volume loops

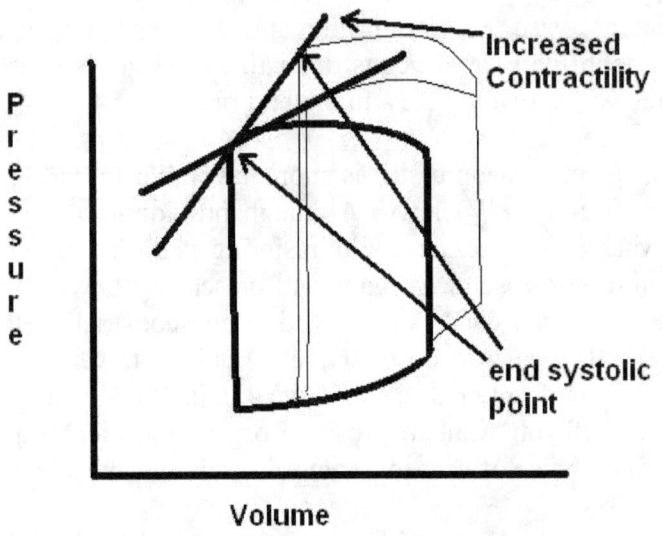

Figure 11.1 Pressure volume relationships and contractility

Connecting the end systolic point on the pressure volume loop for two different preloads will determine the contractility from the slope of this line.

The physical exam finding for contractility is the palpation of the PMI (point of maximum impulse) as an estimate of ejection fraction. The patient should be in the seated position.

The only oral agent that increases contractility is Digoxin. The IV medications include the adrenal hormones sympathomimetic amines and phosphodiesterase inhibitors. These agents can improve symptoms but shorten people's lives. They are only used when vasodilators alone fail to improve the cardiac output.

Contractility is decreased with the use of beta blockers, calcium channel blockers and some anti-arrhythmic medications. This is desirable in treating hypertrophic cardiomyopathy. Decreasing the contractility will decrease the amount of obstruction of the left ventricular outflow. The best agents to do this are beta blockers, Verapamil, and Disopyramide.

Compliance (Lusitropy)

Compliance is the stiffness of the ventricle determined from the slope of the volume pressure relationship. If the ventricle is stiff it will take more pressure to stretch it to a fixed volume. The curve is linear until the volume of the heart reaches the limits of the pericardial sack.

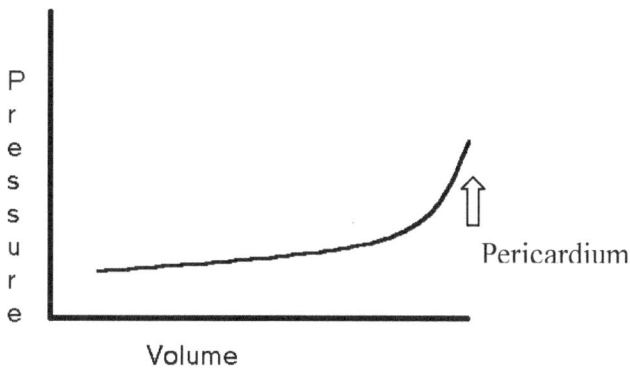

Figure 11.2 Compliance curve

The pericardium, or skin covering the heart, is filled with a small amount of lubricating fluid that allows the heart to move in the sack. If the sack is inflamed, as it is in pericarditis, there is resistance to movement and a great deal of pain. The pericardium is viscous and will stretch if fluid is slowly placed into the sack. If a knife or hairpin is thrust into the heart and removed, blood will quickly fill the pericardial space. The pericardium does not have enough time to accommodate this extra volume and will squeeze the heart. The pressure will rise in all chambers of the heart and prevent further filling. The condition is known as pericardial tamponade and explains why a knife wound to the heart is fatal.

The next illustration explains diastolic heart failure due to the increased stiffness of the heart. The steeper line indicates that for every volume in a non-compliant heart the pressure will be higher. The flat line is normal and it can be seen that increasing the volume of the heart from volume 1 to volume 2 only increases the pressure slightly. The little old lady with hypertension from the nursing home representing the upper curve will have a large pressure change when the volume of the heart is changed the same amount. In fact, the pressure will make the lungs heavy and fluid will pour into the airways causing flash pulmonary edema.

163

Diastolic Heart Failure

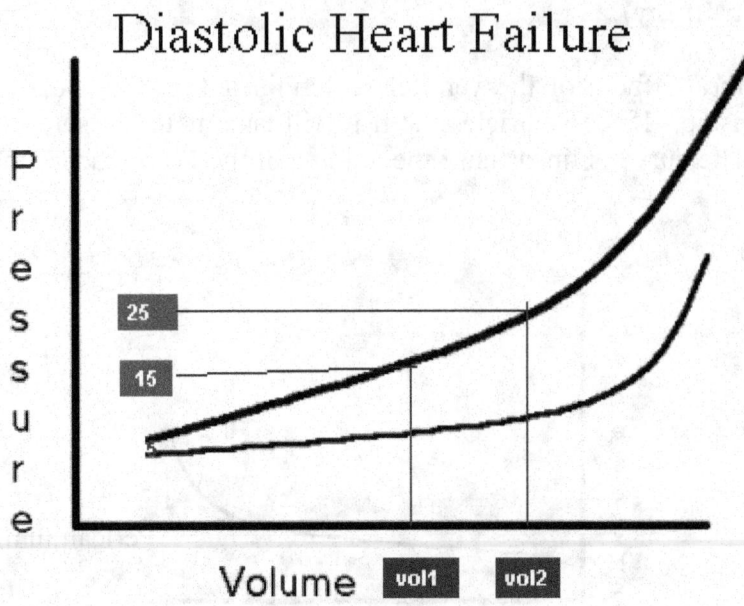

Figure 11.3 Two compliance curves illustrating greater increase in pressure in non-compliant ventricle

The physical exam finding for a stiff non-compliant ventricle is an S_4.

Medications to change compliance are not known since at the time of writing this book no studies in diastolic heart failure have been performed. I suspect that control of blood pressure will be crucial and I believe spironolactone will have a prominent position in the treatment of diastolic heart failure because it is an anti-fibrotic medication. More recently, Ranolazine, the only new anti-anginal agent in 30 years has demonstrated promise as a diastolic heart failure medication. It helps prevent calcium overload in the myofibrils during diastole. Calcium binds Actin to Myosin and is necessary during contraction. If Calcium remains in the vicinity of these fibers some of them will bind and prevent the easy slippage of these fibrils during the diastolic stretch phase. The resistance to slippage requires a higher preload to get the same stretch.

Neuroendocrine

The neuroendocrine system is like God. You can not see it but you have to believe. The medications used to combat the neuroendocrine system are discussed in Chapter 4. Physical exam consists of heart rate and blood pressure. The exam is not very specific other than if the heart rate and blood pressure are too high the neuroendocrine system is not blocked sufficiently. The neuroendocrine system is bad when you have heart failure because it makes congestion worse by holding salt and water in a body already bloated by salt and water. If you are stuck in a desert far from water or injured and bleeding, the neuroendocrine system is your savior.

Geometry and Synchronization

A dilated heart is spherical and an S_3 reflects a volume over-loaded heart. Diuretics can shrink the heart. Surgeons can remove an aneurysm to decrease the volume. They can ring the mitral annulus to combat the spherical dilatation that causes the mitral valve to leak. Heart failure medication can help remodel the heart to a smaller size. The physical exam finding for volume overload is an S_3.

Cardiac dyssynchrony is paradoxical motion of the septum moving in a direction opposite of what it is suppose to move. It is like a V8 engine with two spark plug wires disconnected running rough and inefficient. The physical exam findings of dyssynchrony include a soft S_1 and a paradoxically split S_2. These findings indicate the presence of a left bundle branch block. Cardiac dyssynchrony can be improved with special pacemakers with a left ventricular lead placed near the base of the posterior lateral papillary muscle.

Blood Vessel Properties

The heart is the pump and tissue perfusion is the target. To get to the target the flow generated by the heart must pass through arterial conduits. The properties of these tubes are important for health and can affect the tissues and the function of the heart. The arteries have viscoelastic properties. During systole, the pumping action of the heart, the arteries store energy within the elastic walls so that during diastole the recoil of the vascular wall will maintain flow and keep the pressure from falling. Arterial compliance is defined as a change in dimension with a change in pressure. As we age the vessels become less compliant and are able to store less energy. Diastolic pressure tends to climb and this will make the heart work harder to maintain flow. There are three layers of the arterial wall and each contributes to the physical properties of the artery. Atherosclerosis and calcium add to the stiffness. The physical exam finding is difficult and involves palpating the artery for subjective stiffness. There are fancy machines that use ultrasound, magnetic imaging, and pressure time contours that will give objective results. None of these have been used in clinical practice and remain on the experimental shelf. These devices should be utilized since the stiffness of the arteries is the root cause of suffering. Exercise will improve arterial compliance. If we measure an abnormality in stiffness and start patients on an exercise program and demonstrate improvement of their vessels, these results will provide positive feedback to continue to exercise.

The summary of the hemodynamic performance parameters and the physical exam findings for each are displayed in the following chart.

Parameter	Right Heart	Left Heart
Preload	JVD	S_3 Rales
Afterload	Pulmonary Tap	Carotid Upstroke
Contractility		PMI
Compliance	S_4	S_4
Neuroendocrine	Heart Rate and Blood Pressure	Heart Rate and Blood Pressure
Dys-synchrony		Soft S_1 Paradoxic S_2
Blood vessel compliance		Palpation of Brachial Artery

The measurements of these parameters with a catheter will confirm physical exam findings. The Swan-Ganz catheter was the dream of the doctors whose names are attached to the catheter. They wanted, at the bedside, to sail the catheter through the right heart measuring the pressures and flow within the heart. By knowing the pressures and flows they could optimize cardiac performance. Pressure could be measured through the fluid filled catheter. Flow could be determined by indicator methods using an ice saline injection or by measuring saturation according to Fick. The indicator for the Fick equation is the consumption of oxygen by the body.

The catheter consists of a dual lumen tube with an end hole that could measure pressure at the tip of the catheter. A balloon is proximal to the end hole that can be inflated to help the catheter pass thru the right atrium, right ventricle and pulmonary artery. As the catheter advances into the pulmonary artery eventually it will wedge when the balloon is the same size as the artery. The end hole will now see pressure reflected from the left atrium back thru the pulmonary veins and arteries. The filling pressure of the left heart can be obtained when the balloon is inflated and occludes the pulmonary artery. When the balloon is deflated it will measure the pressure in the pulmonary artery. A proximal port located in the right atrium originates from the second lumen of the catheter. The central venous pressure can be measured from this proximal port.

The pressure is measured by a transducer that converts pressure to an electrical signal. The transducer needs to be calibrated and zeroed to get accurate measurements. The transducer is placed at the level of the right atrium for zero. The measurement can be grossly abnormal if the level of the transducer changes. This frequently can happen if the transducer is not attached to the bed. The catheter is placed at doctor level and then the bed is placed at nursing level. If the transducer is not at the proper level the readings will be false. If you want your patient's blood pressure to appear higher, lower the transducer closer to the floor. This is a common trick of anesthesiologists to get the patient from the operating room to the CCU. (This is a joke. Always keep the transducer at the level of the right atrium.)

Two wires of different thickness are welded together at the tip of the catheter to serve as a thermistor, a tool that measures temperature. Temperature can be recorded by balancing a Wheatstone bridge. This thermistor can be used to measure thermal dilution cardiac outputs, the flow through the heart. Cases using this tool will now be presented using Swan-Gantz hemodynamics and pressure tracing to illustrate cases. We will skip the simple cases of low pressures meaning dehydration and high pressures meaning heart failure.

Case#1
•72 Y/O Male S/P CABG, bleeding – 2 units PRBC given
BP 190/80 Nipride started .1mcg/kg/min with blood pressure falling to 70/60
Arterial tracing shows pulsus paradox

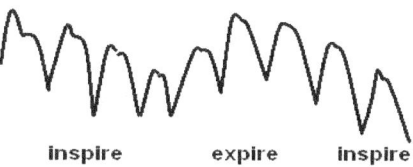

What is wrong? Pulsus Paradox would suggest cardiac tamponade. Pulsus paradox is the inspiratory drop in systolic pressure. Fluid surrounding the heart prevents the heart from filling so if the right heart fills during inspiration the left heart has to get smaller. Should the pericardium be drained?

Examine the initial numbers.

Blood Pressure	190/80
Right Atrium	5
Right Ventricle	25/5
Pulmonary Artery	25/10
PCW	15
Cardiac Output	3.9
Cardiac Index	1.9
SVR	2400

The initial numbers suggest that afterload is high and the index is currently below 2.0. The patient will not survive if the index is not raised above 2.0. Ignoring the clue from the arterial tracing the addition of Nitroprusside causes a new problem of hypotension.

Blood Pressure	190/80	70/60
Right Atrium	5	2
Right Ventricle	25/5	20/2
Pulmonary Artery	25/10	20/8
PCW	15	10
Cardiac Output	3.9	4.5
Cardiac Index	1.9	2.3
SVR	2400	1120

Recognizing the mistake, 2 liters of fluid were given to improve preload.

Blood Pressure	190/80	70/60	120/80
Right Atrium	5	2	7
Right Ventricle	25/5	20/2	25/7
Pulmonary Artery	25/10	20/8	25/10
PCW	15	10	12
Cardiac Output	3.9	4.5	7.0
Cardiac Index	1.9	2.3	3.0
SVR	2400	1120	1063

Pulsus Paradox can also be seen with volume depletion since respiration will determine right and ultimately left heart filling. Hypertension and volume depletion can coexist with high afterload. The patient had high adrenergic tone with intense vaso-constriction causing increased afterload and hypertension. With Nipride the afterload was reduced with resultant hypotension due to preload reduction.

The solution is administering fluids to fill the arterioles and venules decreasing the adrenergic tone. The initial response should have been to give fluids. In this case a fluid administration would have reduced the hypertension to normal. Volume depletion was not realized because the wedge was reasonable at 15. The wedge was elevated because of the intense vasoconstriction. The heart was trying to pump into a brick wall. The main caveat from this illustration is that just because a patient has blood pressure does not mean they are not in trouble.

Case#2
•68 Y/O Female S/P CABG, MVR
After 4 hours blood pressure dwindles with increasing vaso-pressor requirement
120/80 - - 90/75
Urine out put <20 cc's last hour
JVD to 20 cm

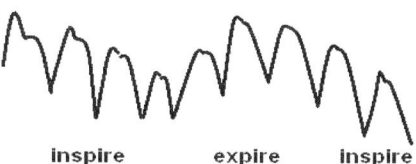

Examination of the neck veins demonstrates a rapid descent and the right atrial tracing demonstrates that is an X descent.

Rapid X descent

Blood Pressure	90/70 HR 130
Right Atrium	20
Right Ventricle	40/20
Pulmonary Artery	40/20
PCW	20
Cardiac Output	3.0
Cardiac Index	1.6
SVR	2133
Pulmonary artery Sat%	48%

What is wrong? The pulsus paradox is seen on arterial tracing. The rapid x descent also suggests cardiac tamponade. The confirmation is the high and equal diastolic pressures. The answer is more fluids until the problem can be fixed by going back to the operating room to relieve the pressure or by milking the mediastinal tubes to allow the bleeding to exit the mediastinum. The electrocardiogram may have demonstrated electrical alternans as an additional clue. The echocardiogram would have demonstrated a small collapsed right ventricle and evidence for fluid around the heart. The hemodynamics can be recorded by observing the fluctuation of the E wave on the mitral vale inflow.

Case # 3
58 Y/O male 4 days S/P MI on Heparin
Platelet count fell from 200K - 96 K
Sudden shortness of breath, hypotension
Positive systolic pulsation pulmonary artery
JVD 14 cm
Kussmal Sign

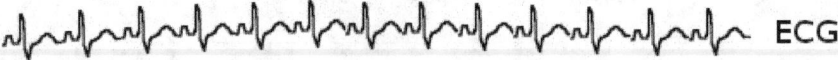

ECG

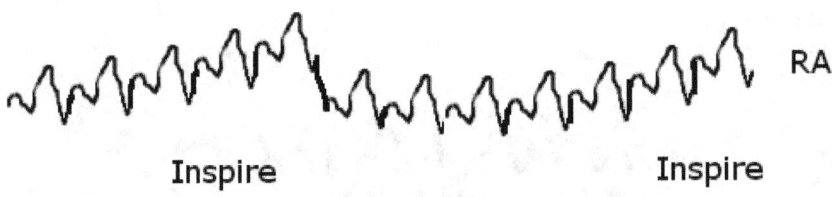

RA

Inspire   Inspire

Blood Pressure	100/50 HR 130
Right Atrium	14
Right Ventricle	40/14
Pulmonary Artery	40/14
PCW	--
Cardiac Output	3.0
Cardiac Index	1.6
SVR	2133
Pulmonary artery Sat%	48%

What is wrong? Is it tamponade? Why can't we get a wedge? The initial clinical scenario suggests that the patient had a platelet drop after being exposed to heparin. The diagnosis suspected is heparin induced thrombocytopenia. Blood clotting is associated with this disorder. Kussmal sign occurs in constrictive pericarditis, RV infarct, or in other conditions that cause the right heart to swell and fill the pericardial space.

In this case HIT (Heparin Induced Thrombocytopenia) is likely and the patient had a PULMONARY EMBOLUS. PCW may be unreliable when there is clot occluding the pulmonary artery. The therapy is to stop Heparin and start (Leech spit) Lepirudin, and Prayer. The right diastolic pressures are elevated because of right heart failure from the high afterload caused by the pulmonary embolus occluding the pulmonary artery.

Case # 4
45 Y/O male with chest pain for 45 minutes, nausea, diaphoresis BP 85/50 P50 JVD 18
Clear lung fields + Kussmal sign 1 – 2 cannon A waves/minute
Describe the ECG

The clues to the ECG include the clinical presentation, which is very consistent with a heart attack. Hypotension and clear lung fields suggest a right ventricular infarction which is confirmed by Kussmal's Sign - the inspiratory rise of venous pressure. Normally, venous pressure falls with inspiration as the lower pressure in the thorax pulls more blood into the right heart from the neck. If the right ventricle is noncompliant because it is infracted the increased flow has no place to go and will be reflected into the neck. Cannon A waves suggest atrial ventricular dissociation or complete heart block. The ECG will demonstrate acute inferior posterior myocardial infarction with elevation in lead V1 and RV4 and complete heart block.

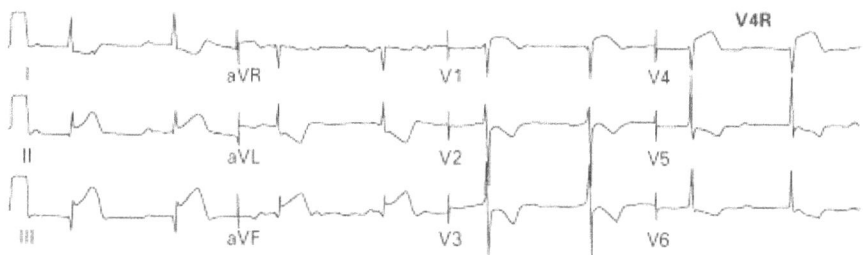

This ECG source was obtained many years ago and the source is lost. My apology

Blood Pressure	85/50
Right Atrium	18
Right Ventricle	30/18
Pulmonary Artery	30/18
PCW	4
Cardiac Output	3.0
Cardiac Index	1.5
SVR	1644
Pulmonary artery Sat%	40%

The initial right heart pressures appear to be consistent with tamponade. The PCW is low because the right ventricle is infarcted and is not pumping blood to the left heart. The hypotension is due to low left heart preload and this can be improved by fluid resuscitation. It may take 10 liters of fluid over the first 24 hours to maintain adequate cardiac output. Kussmal's sign is seen because the right ventricle is infracted and has filled the pericardial space stretching it to its limits.

Case #5

74 Y/O male 5 days after CABG and one day after extubation who develops shortness of breath and requires re-intubation. Initial course after CABG was complicated by hemorrhagic shock.

CXR White Out

Does the chest x-ray represent heart failure or capillary leak from lung damage that occurred after the hemorrhagic shock? Shock can be reversed with blood transfusions, but the damage from the shock will become evident days later. Damage includes swelling of tissues and leaking of fluid. All organs can be affected. Brain Naturetic Protein (BNP) will be elevated in simple heart failure. It may be elevated if the lung damage is causing increase stress on the right heart. The BNP machine is broken. The hemodynamics can be useful in this situation.

Blood Pressure	135/70 HR 100
Right Atrium	12
Right Ventricle	60/12
Pulmonary Artery	60/18
PCW	13
Cardiac Output	7.0
Cardiac Index	3.5
SVR	1040
Pulmonary artery Sat%	70%

The filling pressure of the left heart is normal so the white out is not due to congestive heart failure. Pulmonary infiltrates with low PCW suggest a non-cardiac vascular leak in the Lungs consistent with ARDS (Adult Respiratory Distress Syndrome). The lungs are leaking and it will take time for them to repair themselves. The patient will have to be supported until the lungs again perform their duties.

Case # 6

18 Y/O female in auto accident with chest contussion.
Difficult to oxygenate on FIO_2 75% PEEP 20 cm
No urine output for last 6 hours
CXR – white out

Blood Pressure	100/70 HR 120
Right Atrium	20
Right Ventricle	65/20
Pulmonary Artery	65/18
PCW	24
Cardiac Output	5.5
Cardiac Index	3.0
SVR	1164
Pulmonary artery Sat%	65%

Should we give Lasix?

No, Why Not?

Blood Pressure	100/70 HR 120
Right Atrium	20
Right Ventricle	65/20
Pulmonary Artery	65/18
PCW	24
Cardiac Output	5.5
Cardiac Index	3.0
SVR	1164
Pulmonary artery Sat%	65%

Peep of 20

PEEP is positive pressure that is delivered to the airways to splint the airways open and alveolar sacks that are collapsing due to lung injury. This pressure is delivered into the lungs and affects the pressure in the entire thorax. The heart is in the thorax and the pressure recorded will be changed by the peep. The filling pressure of the left heart is somewhere between 24 and 24-20 = (4) depending on how much of the peep is successfully delivered to the segment of lung that contains the S-G catheter. Her echo pictures demonstrated low cardiac volumes.

The correct answer is fluids not Lasix!!!!

Blood Pressure	100/70 HR 120

Right Atrium	20 – 20 = 0
Right Ventricle	65/20
Pulmonary Artery	65/18
PCW	24 – 20 = 4
Cardiac Output	5.5
Cardiac Index	3.0
SVR	1164
Pulmonary artery Sat%	65%

Peep = 20

When patients have PEEP added, their filling pressure will fall and require fluid. Conversely when PEEP is removed the filling pressure rise and the patient may need Lasix.

The catheter can not be interpreted properly without the knowledge of the patient and the equipment they are connected. Intubation for respiratory failure is commonly accompanied by transient hypotension. This is a result of anesthesia dropping afterload and preload occurs by the bagging. This is a transient drop in pressure but can cascade into a disaster if not quickly reversed.

Case # 7
56 Y/O male with chest pain for 2 hours
BP 90/70 HR 125
Pulmonary edema
S_3

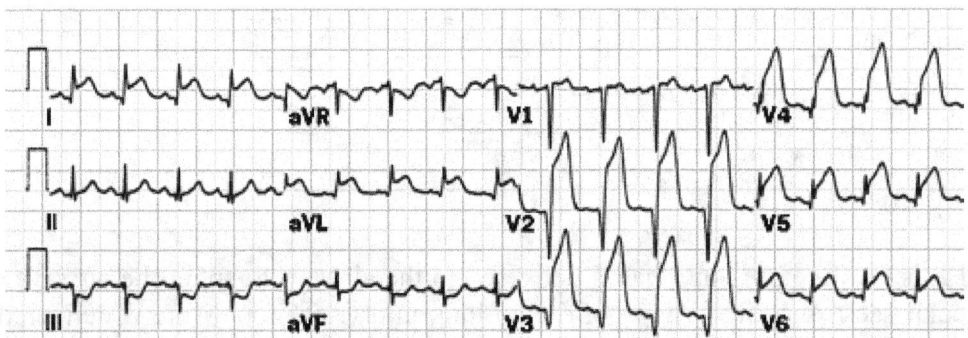

This ECG source was obtained many years ago and the source is lost. My apology

The hemodyanmics follow:

Blood Pressure	90/70 HR125
Right Atrium	20
Right Ventricle	35/20
Pulmonary Artery	35/28
PCW	30
Cardiac Output	3.0
Cardiac Index	1.5
SVR	2024
Pulmonary artery Sat%	45%

What can this very sick patient expect and what can be done to make them better?

The patient has a poor prognosis with a Killip IV MI Classification that implies a 50 to 80% mortality. Reduction of mortality can be accomplished by opening the vessel by the quickest means possible. Thrombolytics can be used for early presentation less than 90-120 minutes. Cath lab is better if immediately available. Balloon pump, intubate to support, Dopamine to increase contractility and Nitroprusside to decrease afterload are measures to improve survival. Ultimately, heart failure medications are titrated as an outpatient.

Case # 8
56 Y/O female with inferior infarct 3 days ago and single vessel disease now present with acute shortness of breath.
She has a new murmur holosystolic in the lateral chest.

Blood Pressure	90/70 HR125
Right Atrium	20
Right Ventricle	35/20
Pulmonary Artery	35/28
PCW	30
Cardiac Output	3.0
Cardiac Index	1.5
SVR	2024
Pulmonary artery Sat%	45%

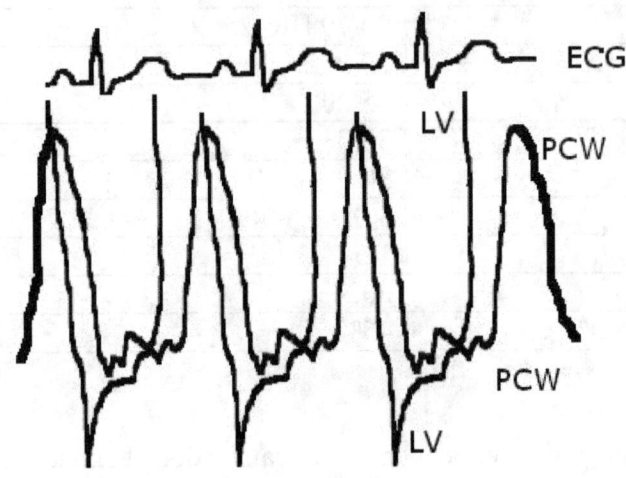

Blood Pressure	90/70 HR125
Right Atrium	20
Right Ventricle	35/20
Pulmonary Artery	35/28
PCW	30 V wave to 50
Cardiac Output	3.0
Cardiac Index	1.5
SVR	2024
Pulmonary artery Sat%	45%

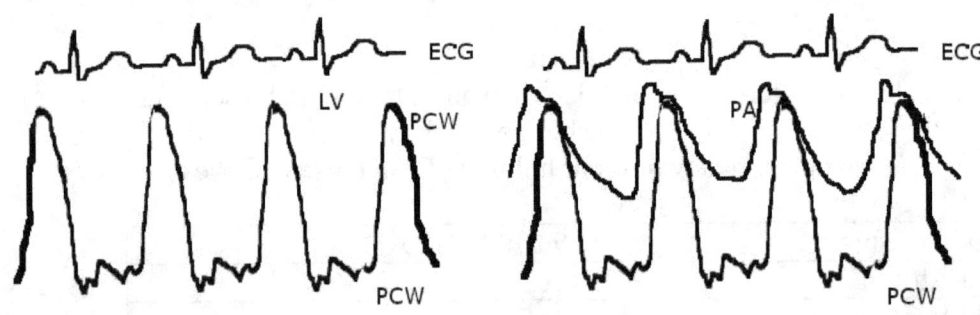

In acute mitral regurgitation, the V wave can be quite large and can be mistaken as a pulmonary artery pressure. In the previous example the V wave is seen in the pulmonary artery tracing. By timing the V wave with the electrocardiogram, you can demonstrate a timing difference to show you are in the wedge position.

How can the wedge position be confirmed?

The wedge position can be confirmed by drawing a blood gas from the distal port with the balloon inflated. If the catheter is in the wedge position the blood that is drawn back will be from the left atrium and should have blood oxygen saturation above 90%.

Conclusion of hemodynamics:
1) Hemodynamics must use the physical exam to confirm suspicions and verify the numbers. In the CCU we treat patients and not numbers.
2) The assessment must not only include the patient, but the equipment the patient is attached to.
3) Finally, you still may not be sure of where your patient is on the performance curve. Some hearts that are sick require very high filling pressures. To determine where the patient is on their performance curve requires an intervention. If the blood pressure is low, the intervention is always volume expansion. Diuresis, and afterload reduction can then be tried to optimize filling pressures and cardiac output. The patient will change over time as they get better or worse and your job is to always make the hemodynamics better.
4) Once the cardiac performance parameters are determined the selection of drugs to improve these parameters can be logically determined.

If this chapter raised excitement, the reader should proceed to the next chapter to see how at least one individual became a cardiologist.

Chapter 12
Cases – Becoming a Physician, Becoming a Cardiologist

The jungle was dampened with smells of thick humidity immersed with floral scents. Tracy's curly hair had been tied back in an attempt to keep the volume of hair, and sticky humidity at a manageable level. Collette, with her straight black hair in pigtails was managing the heat very nicely. In addition to the floral scents, the fresh clean smell that comes just after a rainstorm was permeated by mosquito spray. The foursome was exhausted from the day's journey. Only Murray appeared to be at home with the jungle. The Houckster, Colette, and Tracy looked at each other with anxious and tired expressions. Eager for conversation, Tracy asked Dr. Houck why he had become a physician and if he always wanted to be a physician. She was concerned that the early evening would bring mosquitoes filled with yellow fever, Dengue, Japanese Encephalitis. She was hoping all of his hot CO_2 would attract the mosquitoes to his direction.

The Houckster leaned back against the log and flicked off a beetle that was the size of a small chipmunk. The Houckster stated, "I certainly did not start out wanting to be a physician. I grew up in an exciting time when man was just beginning to venture into space. It seemed there wasn't anything more exciting than getting aboard a flaming rocket and sailing into outer space. I wanted to be an astronaut; however, they had to be perfect physical specimens and certainly needed to have good eyesight. My eyesight was nothing without spectacles. I had an astigmatism that would always make it difficult to paint fence slats let alone distinguish parallel lines on monitors. I told my girlfriend in 11th grade that I had a grand plan to become an engineer, then a physician. Perhaps with these two skills mastered, NASA would consider me for astronaut training. I'm not sure that I actually believed what I told my girlfriend. I really just wanted her to like me little bit more."

Life takes many turns and interests change as one experiences the joys and burdens of becoming an adult. Practicality oftentimes wins over dreams. I enjoyed my engineering training and especially enjoyed trying to figure out how things worked. I still had a dream of becoming an astronaut, so I did take one course in the biological sciences that would fulfill requirements for entering medical school. I chose a biophysics course. There were no pre-med students in this course, since it may have been difficult to obtain an 'A.' I really wasn't very sure about medicine and thought that I would be squeamish working around corpses or getting anywhere near blood products. I decided I would continue training as a biomedical engineer. I soon learned that a biomedical engineer could do very little without a champion known as a physician. The physician would take the responsibility of having a device approved and experiments conducted on patients. A biomedical engineer could only come up with the ideas and then sit by the wayside as the physician performed his duties and took the credit. The original goal to become an astronaut by becoming a physician had now been replaced the intense training to become a physician.

I was an older medical student, 26 years of age, in a class that had some students who were barely 19 years of age. I had the advantage that the Air Force was supporting my training and compared to other students I was filthy rich. I was not incurring any debts and had meager living expenses that could pay for my apartment and most of my food. Entertainment was provided by conversations with my new bride. My wife and I would entertain frequently because we were the only ones who could afford the beer. Medical school was not really a pleasant experience. The amount of material taught was enormous and for one with poor memory it would take hours sitting isolated in a room studying to retain the volumes of data. My only outlet was Farrah Fawcett in a red bathing suit poster pinned to the wall directly in front of my card table desk.

Drawing blood for the first time causing patients pain and anguish, placing NG tubes, performing rectal examinations were all part of becoming a physician. I realize now that those skills were not very important and certainly can be learned with repetition. At the time, a single failure would make one want to reconsider further training. With the help of a supportive carpool, a note taking service, and very supportive wife, I graduated. The entire time I spent in medical school was simply to survive. Becoming a physician and what it means to actually be a physician never registered.

Some students with all A's had the ability to always come up with the correct answer when the test questions were worded properly. Some of the same brilliant students, however, lack the skills to obtain the history from the patient. Patients do not read textbooks. Textbooks are boring. Patients may express discomfort or feelings in ways that are foreign to us. A good physician is one who listens to the patient but also listens between the comments. The physician watches the patient, observes the patient and the family during interview. His suspicion and differential diagnosis is formulated. He becomes Sherlock Holmes. The interview is directed towards confirming or refuting a diagnosis. Patients have to be allowed to answer the questions in their own manner. Patients have to be directed in order to get an answer within the allotted 20 minutes. There is an art of sitting back and only listening at times and at other times leading a rambling patient. The history and extraction of complaints from the patient is still the most valuable technology in the practice of medicine. If I don't know the answer within the first three minutes of the interview, it is unlikely that a disease specific diagnosis will be obtained.

Training after medical school is like an apprenticeship. It involves learning on the job and observations of what your staff and fellow residents do in order to heal patients. The science and analytical skills we practiced in medical school is now used on every patient encounter. Training and practice in medicine never ends. The term practice of medicine sums up a career in medicine. The doctor is always striving to become better but never achieves perfection. Mentors are fellow residents, interns, physicians, nurses, and some patients who shape the physician as he is trained. Training at the Air Force's premier hospital with the top 10% of the medical school graduates was a benefit which I did not realize while I was at Wilford Hall. We were constantly under fire by our senior residents and chief residents. Any demonstration of lack of knowledge, any weakness, would be an opportunity for further reading assignments. We had fun during our training, but acquiring knowledge and making patients better was the primary focus.

When I became a senior resident I adopted the techniques of my predecessors. I felt that I was a kinder, gentler senior resident. I would start the morning by asking a question of the resident. If the resident did not know the answer he or she would be thrown up against the wall and he or she would go read the rest of the afternoon. This routine would be repeated daily asking a question and then throwing the resident into the wall. On the fourth day a question would not even be asked. The resident would be thrown against the wall for no reason and he or she would read the rest of the afternoon. Medicine is gentler in today's environment. Although, one can no longer throw a resident against the wall, similar psychological torture can be accomplished. Reading about the patient's illness helps to cement the knowledge for future use. A good resident is not necessarily a brilliant resident. A good resident is one who always wishes to learn more. A good resident always makes sure that the history, physical examination, and laboratory are consistent with a common diagnosis and that all abnormalities are explained. A good resident, "does not trust their mother" and only accepts data that they have viewed themselves. A good resident always checks the lab that was ordered and has a plan for the results. A good resident is a pessimistic resident who always expects the worst from their patients and by expecting the worst is prepared to prevent these dire imagined outcomes.

The training physician picks a physician mentor. They cannot be assigned. The training resident has to find something of the mentor that he desires to become. I had many mentors during my training and still look for mentors. Things that I can learn from these expert physicians will make me a better physician. Training and practice of medicine never ends. Some of my mentors have been scary. My chief resident was a small wrestler with the nickname of Mad Dog. He had great knowledge and the ability to find weakness in any of his residents. During morning report he would never hesitate to use this weakness to turn a large resident into a quivering bowl of Jell-O. His technique was effective and his residents came to morning report much better prepared knowing that a battle would ensue. He was successful in eliminating quite a few residents so that in my third year I was on call as much as the interns. More call was more of an opportunity to learn.

A red-haired Air Force Academy graduate was able to teach me history and physical examination. More importantly, he instructed me in specific patient management using the personality of the patient to pick the proper therapy. He was the influence that directed me towards cardiology. I wanted to be able to diagnose and treat patients as effectively as he could. He had a sense of those patients who would benefit from aggressive interventions and a better sense for those who would not. The most difficult decision in cardiology is to do nothing. Doing nothing can be criticized but may be the best course of action if the little voice within your brain tells you the patient is more likely to survive by doing nothing. A retired Army Colonel who had a stethoscope calibrated to Proctor Harvey's stethoscope taught auscultation of patients to me. He felt the length of the tubing supplied by the manufacture was too long. The shorter the tubing the better the auscultory findings, however the closer you had to be to the patient. Some patients are not pleasant to be near. He measured his stethoscope to Proctor Harvey's and I measured mine to his. I still have mentors and am looking for new mentors. At the

American College of Cardiology national meetings there is opportunity to interact with learned professors and mentors. I especially enjoy the clinicians who also have an interest in science. A current mentor, Clyde Yancey, is a Heart Failure Specialist in Dallas, Texas.

At my present job I try to emulate a semi-retired physician Charlie Welch. Charlie has a great life. At 75, he teaches the art of ECG interpretation to medical students and residents in the morning and reads EKG's until noon. He then has time for cards, his wife, New York Times, and relaxation in the afternoon. I inherited many of his patients and found it very difficult to gain their confidence after his excellent care. Charlie is truly loved by everyone and I suspect he has sore ribs from all of the hello hugs he encounters. He and I have a lot in common including previous military service, Northwestern University, and love of sports cars. Charlie is now driving a silent environmentally friendly automobile that scares the heck out of me in the mornings as he drifts silently into the parking space. Charlie has taught me that a smile will greet the world in a more positive manner. Even when he had difficulty swallowing from a terrible diagnosis, he continued to smile and go about his daily business. I want to greet the world and be as positive as Charlie.

A good physician is one who practices within the guidelines. Evidence based medicine is a very safe practice. Evidence based medicine may be similar to snobbery because only certain patients are treated with certain named brand drugs. Most patients do not fall into the proper patient groups and even evidence-based physicians have to modify their treatments to the individual. The great physician is one who practices ahead of the guidelines. I like to present several cases where practicing beyond the guidelines became a necessity for these patients.

Case # 1 Plasmapheresis for myocardial Infarction

The first case illustrates a teaching point all residents need to learn: what to do when you don't know what to do? Patients can present with terrible disease processes with expected bad outcomes. There is no fear when one can make the diagnosis and can expect the worse. The challenge is to prevent the worst. The case I'm presenting is one that would instill fear into any cardiologist. The fear is observing something that has never been seen before. The fear in this case is heightened because it involved a 31 year-old female veterinarian who gave birth just three months prior. Her problem began the day before Halloween when she was attending a dinner seminar. During the dinner, she developed chest discomfort with radiation to the left shoulder. She became quite uneasy and felt she had to get up and move about the room. It took nearly 15 minutes for this discomfort to become relieved and she was able to finish her meal.

The next day she went to her primary-care physician's office because she had had recurrence of the same feeling again with eating. A thorough history and physical and even electrocardiogram were obtained during this visit. The electrocardiogram was compared to old tracing and demonstrated no new findings. Obtaining an electrocardiogram for current symptoms and age was fortuitous but unrevealing. The primary care physician must have felt "the force" and sensed an impending problem. An expert EKG reader always compares to the old tracing and this was performed.

The physician explained this as a new mother syndrome. Claiming these symptoms were likely dyspepsia due to a stress, she was prescribed several agents for current life stresses and dyspepsia.

The next day, the pain unfortunately continued and this time occurred without having any relationship to food. She did try to relieve the discomfort with Zantac and was unsuccessful. She went to the emergency room because of the pain and foreboding feeling. The emergency room physician noted that she was 12 weeks postpartum, that she had a reflux type of pain for the last several days. The pain did not worsen with exertion, yet it was sharp and continued to radiate to the center of the chest. Initially it was associated with eating, but now was there constantly. She's also had numbness in the left arm and neck discomfort. She did not complain of any shortness of breath with the pain. The pain did get worse with deep inspirations.

During the course of the patient's workup, the pain became much more severe. The patient's husband was quite concerned about his wife's discomfort and discussed these issues with emergency room physician. In an effort to control the patient's pain, she was given intravenous 1/2ml of Dilaudid. During the injection, the monitor demonstrated new ST segment elevation, bradycardia, PVC's, and R on T phenomenon that resulted in ventricular fibrillation. The young, freckled, green eyed, strawberry-blonde 31 year-old female, a beautiful new mother, had suddenly died of cardiac arrest. She was promptly resuscitated with electric shock to the chest. Repeat electrocardiogram demonstrated transient ST elevation in the lateral leads consistent with an acute lateral myocardial infarction. Within 30 minutes these changes resolved and the electrocardiogram was again normal.

Cardiology was called to see the patient because of the unexpected cardiac arrest. Bedside echocardiogram demonstrated mild hypokinesis, weakness of the posterior lateral segment. The patient was now pain-free and the electrocardiogram returned to normal. The history was again obtained, and was more consistent with coronary artery spasm which is a spontaneous self-squeezing of the coronary arteries as opposed to atherosclerotic blockage related illness. The patient had no pre-existing exertional chest pain. Her past medical history included gastric bypass surgery with a 100-pound weight loss and a recent pregnancy 12 weeks prior and hypertension induced by pregnancy. She was previously obese but was now at a healthy150 pounds. Her family history was negative for any sudden cardiac death or coronary disease at a young age. She had no allergies. Her medications included recent refill of propranolol for her hypertension, ibuprofen 800 mg twice a day, birth-control pills, multiple vitamins, Celexa, and Flexeril both of which had been started the previous day. Her physical examination demonstrated blood pressure 140/80 normal saturations, a pulse of 76, regular. The only abnormality was an S4 on examination, which would suggest possible myocardial stiffness from ischemia. The chest x-ray demonstrated normal cardiac silhouette.

The assessment at this time was ischemic ventricular fibrillation, sudden cardiac death, from R on T. The etiology of ischemia could be secondary to a ruptured or eroded atherosclerotic plaque. However, the patient's pain pattern over the last several days was more consistent with possible coronary artery spasm. The patient's bedside echocardiogram demonstrated posterior lateral hypokinesis after her resuscitation. The wall appeared to have adequate thickening but weakness in the distribution of the circumflex coronary artery.

In view of the patient being pain-free with normal electrocardiogram and resolution of ST segments she was placed on medical therapy with heparin, IIb IIIa platelet inhibitor, intravenous lidocaine and intravenous nitroglycerin. She was given a Statin medication, and ACE inhibitor. The patient was consented for cardiac catheterization with planned elective catheterization the next morning. The purpose of the catheterization was to determine the patient's coronary anatomy and to decide if it was atherosclerotic versus vasospastic disease of the blood vessels.

The patient did well overnight with several brief occurrences of pain. Other than a brief increase in the nitroglycerin drip, further therapy was not required. The patient's cardiac enzymes were mildly elevated next morning but the electrocardiogram demonstrated no evolution and remained normal. The patient was sent to cardiac catheterization where during her angiogram a hushed silence came across the room because of the findings. All three of her blood vessels had demonstrated severe blockage in the distal portion. The right coronary artery, the circumflex, and the left anterior descending demonstrated blockages that could not be repaired by angioplasty and were not suitable for bypass grafting due to the diffuse severe nature of the disease. This finding was a surprise and the best description of the patient's coronary anatomy was similar to what one can see in cardiac transplant vasculopathy due to rejection vasculitis of the coronary arteries. Because of the diffuse nature for disease process she was loaded on Plavix with the hope of stabilizing her recurrent infarction. Throughout the next 24 hours, she had recurrent pain with EKG evidence of the inferior infarct followed by an anterior infarct. The senior resident on the cardiology service stated, "Dr. Houck, you have to do something she is infarcting her entire heart."

What to do when you don't know what to do?

Fear is not a useful emotion. Fear is the foremost emotion and was attempting to take control of my cerebral cortex. Fear can prompt one to action without forethought. Fear must be conquered. I was thinking. The first issue was to keep this young mother alive to raise her child. An artificial heart would be required to support her circulation if she does indeed infarct all three blood vessels. Plavix would be a problem for 5 days and would need to be stopped and IIb IIIa's restarted. The dilemma was to keep the blood vessels from clotting and still allow a surgical procedure to place a left ventricular assist device (artificial heart). Calls to transplant centers were initiated as well as calls to our surgeon to help us with assist devices. Our surgeon knew that the assist devices in our institution were too large and balky for this patient and suggested discussing the case with a transplant friend in Houston Texas. The plan was to transfer the patient to an institution where a Jarvik pump could be placed to support her circulation while her clinical course was further investigated. If she turned for the worse a heart transplant might be required. The Houston Physician did have a suggestion when he heard that the coronary arteries looked like a transplant vasculopathy. He was a transplant doctor and heard of unpublished cases of postpartum patients who had similar findings. The rumor was that they might respond to plasmapheresis and suppression of the immune system similar to heart transplant patients.

That very night the central line was placed into her femoral vein on full anticoagulation and she was started on steroids and plasmapheresis in the setting of multi-vessel acute myocardial infarction. She demonstrated other findings during our course of treatment including pericarditis and valvulitis with the new development of mitral regurgitation. She had no skin rashes or any other evidence of immunity related disease process.

There were no guidelines for the treatment of this patient. Instead, only a helpful suggestion based on a description of a process that appeared in patients who had previous transplants. We were lucky; in our group of doctors we have our own health maintenance organization, which the patient was a member. As physicians, of our own health maintenance organization, we never have to ask for permission to treat patients since we are duty-bound to provide the best care. The belief is that the best care will be the most economically feasible care. If this patient had other insurance, it is unlikely that plasmapheresis would have been approved for her care.

The team assembled and walked into the patient's room. She certainly did not look like a dying patient. She was quite attractive with her hair pulled back and she had even placed the rudiments of makeup. Makeup is a positive prognostic sign. She was no longer in distress and was breathing fine. She was calmly reading a book entitled "Kill as Few Patients as Possible: and Fifty-Six other Essays on How to be the World's Best Doctor" by Oscar London. Attempting to ignore the book that was in the patient's lap, I began summarizing her problems that included a coronary angiogram, which demonstrated findings that I had never seen before. We felt that this likely represented a vasculitis probably related to her previous pregnancy. If this was garden-variety coronary artery disease, the therapy could only be medicine since the vessels were not suitable for bypass grafting or for angioplasty.

We explained to this young educated mother that we felt that she had a disease condition that we did not fully understand. We discussed her case with a number of different experts who also did not have a clue. We had found one individual who suggested that treatment with plasmapheresis for an immune related attack on her coronary arteries could be beneficial. After examining all the options we felt that this was the best one to proceed with and were recommending that she be treated with steroids, plasmapheresis, and immunoglobulin. We could not assure her that this treatment would work. We would continue her standard care for myocardial infarction. We told her we had contacted transplant centers to help us if she should deteriorate. The patient had a medical background and was very familiar with immune related diseases. It took less than 30 seconds for her to understand the gravity of her situation and this really represented her only treatment option. With a confidant smile she said, "sure, let's get started."

Steroids, plasmapheresis three times the first week, and intravenous immune globulin given between treatments were initiated. The patient demonstrated anterior apical infarct on EKG and on echo. The apex was dyskinetic with risk of forming thrombus. The posterior lateral segment was extremely weak with evidence of EKG and echo findings of infarction. The inferior wall showed acute ST elevation but the wall motion in that segment was preserved. The mitral valve was leaking, and the pericardium was inflamed. Multi-vessel infarction had been described before but never to the extent exhibited in this poor young mother. During the course of her therapy she improved with less chest pain and improvement in her EKG and Echo. She did not lose R wave forces. She was discharged home with a defibrillator and a renal dialysis catheter placed into the neck for her plasmapheresis. She had another episode of chest pain while at home and again was admitted to the hospital. The pain was shorter and the elevation in cardiac enzymes was less. She was treated with increased steroids and additional plasmapheresis and the pain resolved.

One week after her second discharge she was shopping in the Dollar Store and received 6 mule kick electric shocks to the chest. My heart sunk when the ER notified me of her change in status. I thought we were out of the woods with a miraculous recovery from an unknown disease. I was so nervous about this patient that I discussed her unusual situation with my parents who I still depend for advice. My parents put her on their prayer list and she had immediately improved. What could have gone wrong? The defibrillator had been placed because of the inability to correct her ischemia with high likelihood of recurrent ventricular fibrillation. She needed this protection as a bridge to a possible transplantation. If she had recurrent ischemia what would be our treatment options?

In the ER she sat with a curious smile and was not in distress. How can a patient be so composed after six conscious shocks to the chest? I was thankful she was awake because I knew the shocks were likely inappropriate. We use anesthesia to put people to sleep before we shock them. She was as calm as the water on a lake in the early morning just as the sun is rising. A quick evaluation of the pacemaker and a chest X-ray demonstrated lead dislodgment. She did not have recurrence of her ischemia; she had false shocks due to lead dislodgement. I breathed a sigh of relief. Four months later she was completely pain-free back to her veterinary duties and raising children. She had no evidence of myocardial injury on her echo and a 64 slice CT scan of her heart demonstrated no evidence of the severe coronary artery obstructions that she demonstrated previously.

Peripartum cardiomyopathy is a disease process that occurs in young women and causes significant left ventricular dysfunction. Some of these women will spontaneously improve others may go on and require transplantation. Cardiology does not know how to treat this illness. Immune suppression has been tried without much success. This patient may have had a variant of postpartum cardiomyopathy. The patient's pain pattern prompted her angiographic evaluation. If she had not had pain she likely would not have gone to angiography and her coronary arteries would not have been treated with plasmapheresis. This may not be a new disease process but a window into the cause of postpartum cardiomyopathy. Although it is preliminary to suggest that a trial should be undertaken to treat postpartum cardiomyopathy with plasmapheresis, other more dangerous therapies have been tried. The concept may apply to treatment of all myocardial infarctions in an effort to keep inflammation from permanently scarring the heart. The interaction between the mother and fetus with the possibility of fetal cells circulating within the mother's system may cause an immune reaction that may attack the myocardial cells as an innocent bystander. Removing the offending antibody by plasmapheresis may be able to reverse the immune reaction. Dying myocardial tissue also causes an increase in inflammation that may permanently injure the heart. Perhaps plasmapheresis would be good for the common myocardial infarction that affects millions of men and women yearly.

Could gastric bypass be a contributor to this disease? Will we see this in the thousands of obese patients who had this therapy? The gastrointestinal system is an immune modulator. The GI track sees the most foreign substances and mounts the first line of defense against bacterial and viral agents. Does gastric bypass modify this response and make those patients more susceptible to immune related diseases?

There were no guidelines to direct this patient's care. It was felt that plasmapheresis had a favorable risk to benefit ratio and demonstrated better than expected results. Were we lucky or, did prayer solve the problem? An individual case can help develop a hypothesis, however, detailed and lengthy study needs to be applied to prove our therapy is good. Plasmapheresis may become standard care in the treatment of post partum cardiomyopathy or for myocardial infarction. It will take time and study to prove its benefit. I believe it deserves rapid study.

Case # 2 Right Ventricular Infarction
And Papillary Muscle Rupture

Beep. Beep. Beep. The page from the Emergency Room was received in the cath lab while emergency angioplasty was being performed on the first myocardial infarction of the night. The news was not good. An 80 year-old female had just been intubated for respiratory failure and pulmonary edema. Her blood pressure was 90 systolic while on dopamine. Her story began several days prior with new onset of gout. She was prescribed colchicine for the painful joint and had developed dehydrating diarrhea that was profuse. Early in the morning she had chest pain and had become progressively more short of breath and eventually agreed to come to the emergency room when she was near death. The electrocardiogram demonstrated inferior Q waves and ST abnormalities. Her laboratory was even worse with a creatinine of 4 and a bun of 100. The force was strong that night. With only a 2 minute communication from the emergency room doctor, the Houckster had a sense that the unfortunate female had a ruptured papillary muscle. It made sense she should be in shock. However, it did not make sense she should be in pulmonary edema with a chest X-ray that demonstrated white out. The Houckster sent the senior resident to the patient's bedside to get more history and to listen for a pistol shot murmur in the left axilla.

The resident reported that he did confirm the murmur. He reviewed her history with the family and electronic record and determined the elderly female was in great health until the flare of gout. She had normal kidney function on prior evaluation and was on no chronic medications. Her mind was sharp and still ruled the family as evident by the 7 family members present in the family consultation room. She was nearly 18 hours from the onset of her inferior infarction.

Blood pressure was 90 – 100/60 (on dopamine) with a heart rate of 125. Physical examination revealed dilated neck veins and scalp veins with Kussmal's sign indicative of right ventricular infarction. The carotid upstroke was brisk but of low volume, systolic pulsation of the pulmonary artery was present. The PMI was non-displaced and discrete indicating a good ejection fraction. Auscultation of the heart revealed a soft S_1 and a pistol shot mitral regurgitation murmur. The lungs demonstrated crackles.

The electrocardiogram demonstrated inferior Q waves and a tall R wave in V1 with ST elevation in V1 suggesting right ventricular infarction.

The fellow rolled the echo machine into the room and demonstrated a ball of meat flailing into the small left atrium. This was the 'force' predicted ruptured papillary muscle which had become detached. The right ventricle was dilated and akinetic and the left ventricle had an ejection fraction of 50% or greater with a small inferior infarction.

It was time to make a decision. The family would help decide, but it was the attending physician who had to decide between quiet room, or aggressive therapy gently guiding the family to the same decision. It was an unknown area. He had treated many right ventricular infarctions and multiple papillary muscle ruptures. He had never treated both conditions in the same patient. A recent article on the benefits of balloon pumping and cardiogenic shock would suggest a period of stabilization on the balloon pump would benefit some patients. If the right ventricle could be supported until recovery the patient could undergo surgery to replace the mitral valve. The decision maker occurred while discussing options with the family. The odds were against the patient. She was in previously good health and had a zest for life. Her brain was still functioning.

She went for a balloon pump and limited angiography with less than 30 ccs of dye demonstrating a closed right and 70% left anterior descending obstruction. A Swan-Ganz catheter was placed that surprisingly demonstrated no V wave on the pulmonary capillary wedge pressure tracing. This was surprising since all the textbooks site a giant V wave due to papillary muscle rupture. To generate a V wave the left atrium has to be non-compliant so that the regurgitant volume will quickly raise the pressure. In our patient, the left atrium was small because the infarcted right ventricle could not pump blood through the lungs to fill the left atrium. An under filled left atrium is very compliant and thus did not generate a V wave. The right coronary was left closed. It could have been opened. The extra dye could have caused further damage to the kidney. The vessel was closed for a considerable length of time making myocardial recovery unlikely. The patient if she were to survive would eventually need surgery. The right ventricle may recover quicker if the vessel was opened was the only pro argument. The benefit was less than the risk.

After the balloon pump was placed, the patient was not much better. She still had blood pressure of 90 and heart rate of 115. The dopamine was discontinued. Vasopressin was added and she was transfused 2 units of blood despite her seemingly adequate hemoglobin of 11.1. The extra preload from the blood and organ sparing pressor effect of vasopressin gave a sign of encouragement. The urine was flowing at a meager 30 to 50 cc's per hour. This was incredible considering the creatinine was 4. Even this value had improved suggesting that cardiac output was adequate. If the 10 liters of saline had been given the patient would not have been improved but would have looked like a puffer toad bloated with the face being unrecognizable. All the edema would eventually resolve if the patient would survive.

The official echo was performed the next morning with the surgeon at the bedside. The right ventricle was not moving and dilated. The papillary muscle was still flailing into the left atrium. The surgeon wisely stated the unfortunate patient was not ready for an operation and he should be contacted if the right ventricle regains function. A discussion around the bedside turned to Viagra. This medication was contraindicated in the setting of myocardial infarction. Before nitroglycerine is administered the emergency room doctors have been trained to ask if the patient had taken Viagra within the last 24 hours. They would always have to add to the statement that their life depended on how well they remembered. Male patients do not volunteer Viagra information and need strong coaxing. The reason Viagra is dangerous, is that nitroglycerine will have super potency and if it is given with Viagra blood pressure will plummet and stay low for an extended period of time. The Houckster had prescribed Viagra previously in an unusual case of chronic RV infarction. He was aware that it could be used even in low blood pressure. He knew of its reported benefit in pulmonary hypertension. The surgeon volunteered that he had attended a conference where some smart but crazy Italians had reported the use of Viagra in right ventricular infarction. A quick literature search yielded no hits.

A joint decision was to try Viagra. Through the NG tube 12.5 mg of the male potency drug was given with a gentle drop in pressure and an amazing increase in urine output. Two hours later the blood pressure was improved. During each 8 hours the dose was increased. The right ventricle demonstrated decreased swelling and improved contractility. Within three days the creatinine had normalized, blood pressure was 130 with a heart rate of 90. Vasopressin had been discontinued and Natrecor was initiated. The lungs had cleared. It appeared a window of opportunity was present to get the valve replaced. A window is difficult to judge but is clearly recognized when it closes.

The patient went to the operating room and had a difficult time. The surgery was flawless but the right ventricle is very sensitive to insult and is the limiting factor in the success of heart surgery. She came out of the operating room in shock with an open chest because the right ventricle had failed. She was on standard multiple drips with intense acidosis and no urine output. In the heart room of the coronary care unit she was placed back on vasopressin, Viagra, given more blood for preload and weaned off of her multiple drips. The acid cleared and the kidney function again improved. This almost never happens in patients over 80 because they have underlying kidney dysfunction. After 2 days she was felt stable to have her chest closed. The struggle to keep the head of the family alive now came to an unfortunate end. As the dressing was removed to expose the heart the entire front of the right ventricle tore from the heart causing near instant loss of blood in an un-repairable condition. This catastrophic event ended her struggle. The family was thankful for the efforts of the multiple health care providers.

Case # 3 Pacemaker for Mitral Regurgitation and LBBB

Thelma was an 81 year-old female who had been well for most of her life. She had an ECG performed more than 25 years ago that demonstrated Left Bundle Branch Block. She did not feel any different after she had the ECG and lived without misery for the next 25 years. Slowly she began to have exercise intolerance and began to depend on family members for shopping and house chores. She began to have swelling in her feet and more shortness of breath with activity. One night she awoke with shortness of breath and called family members who found her in a confused state with bubbling respirations. EMS arrived and administered oxygen with improvement of her confusion. She was carted to the hospital where standard heart failure therapy was administrated.

She had an Echocardiogram that demonstrated severe mitral regurgitation and ejection fraction of 45%. She was discharged and sent to the heart failure clinic where her medications were adjusted. Despite her titration on medications she was readmitted to the hospital, and eventually it was concluded she should undergo valve replacement surgery to control her heart failure symptoms.

The heart failure nurses referred Thelma to the Houckster to make his recommendations. After looking at the echocardiogram he could see nothing wrong with the mitral valve. It leaked severely but had normal thickness and no torn chordae to explain the severe mitral regurgitation. He did note the paradoxical septal motion seen with left bundle branch block and remembered how the mitral regurgitation would lessen after a new therapy of bi-ventricular pacing. This therapy was only for left ventricular dysfunction with an ejection fraction of 30% or less. She did not fit into this category. He estimated valve replacement surgery would be 5% fatal in this patient and that her heart failure symptom may not get better. They should not get worse but in some patients after mitral valve surgery the ejection fraction would plummet. It is very difficult to sense which patient will have a bad outcome. The Houckster had this feeling about Thelma. He wanted to push medical therapy farther and really wanted to place a pacemaker to cure the root cause of Thelma's problem cardiac dys-synchrony. The Houckster had no idea why it had taken 25 years for the heart to remodel in an adverse manner. He did not understand what might have changed in the elderly female's heart. He was confident that restoring cardiac synchrony would improve Thelma and save her from a potentially deadly operation.

At the next visit to the Houckster, Thelma was no better. She had not been in the hospital, but was still short of breath with simple activities. Her diuretics were adjusted and just as she was leaving the room, she volunteered that she had a pass out spell three nights prior. She was on the way to the bathroom and woke up on the floor. Now the pacemaker plan could be put into place.

The first order of business was to convince the Electrophysiologist that a third lead and resynchronization therapy was justified despite the good ejection fraction. This was not a simple task since the insurance company may deny the claim and the hospital would be stuck with the difference. Thelma was not of any help because during her consultation with the EP doctors she had forgotten about the pass out spell. The EP doctor must have had the same bad feeling about valve replacement and agreed to do the procedure.

The pacemaker placement went well, except the third lead could not be placed because Thelma did not have a cardiac vein in an acceptable position. She would have to see the surgeon who agreed to replace her valve for a thorascopic placement of the third lead. The surgeon was gracious but said it would not work but he would be happy to place the lead and then do surgery on her valve later. The Houckster said a "thank you" and reminded the surgeon to place the lead at the base of the papillary muscle so the muscle would be the first part of the heart to contract slamming shut the mitral valve.

The surgery went well but she was admitted 3 days later with pulmonary edema. She was diuresed and sent back to the Houckster to sort out her problems. Thelma had a new problem. She had complex ventricular ectopy and was in bigeminiy or trigemiy all of the time. The ectopy meant she was not synchronous. She was started on Amiodarone to control the ectopy. One month later she had all of her diuretics discontinued. Her heart failure hormone had fallen from 900 to 90 and her murmur had disappeared from her physical exam. One year later no murmur could be auscultated and breathlessness remained only a distant bad memory.

14 Year Old Heart Attack

The pediatric intensivist contacted the adult cardiologist on call to do a late night echocardiogram. The young man had been having pass out spells for the pass 6 months. He was being treated for a seizure disorder. He had been horsing around at school and with exercise had chest pain and passed out again. The Intensivist was suspicious that the heart was the cause of passing out and not the seizures. The child arrived in good condition without complaints. He did have transient EKG changes noted on the outside emergency room tracing. The Echo demonstrated wall motion abnormality in the posterior lateral segment consistent with ischemia or myocarditis. Heparin and Integrilin were started just as in adults with a plan for an elective catheterization the next day. The elective portion of the cath changed when the young man got out of bed to go to the bathroom and again collapsed. This time he awoke in fulminate pulmonary edema and was dying.

The Cardiologist felt he had been working too many years when the rare events tended to become more frequent. The young man had an anomalous left main coronary artery that was being pinched because it ran between the aorta and the pulmonary artery. The main artery was a slit and a vicious cycle was now working. The left main would cause ischemia that made the heart more hypertrophied and stiff. The resulting diastolic dysfunction would raise the pressure in the pulmonary artery causing them to enlarge and further pinch the left main more causing more ischemia. This process is a vicious cycle and ultimately results in death.

Natrecor, intubation with mechanical ventilation, an intra-aortic balloon pump and Lasix resuscitated the boy. An experienced cath lab nurse found a catheter that would fit into the anomalous left coronary that was coming off of the posterior right cusp. It was an All Right Catheter. This demonstrated the problem so the solution of bypass surgery could be performed.

The timing of surgery was critical. The current ejection fraction was 10% after the ischemic event. The EKG had returned to normal and hemodynamics improved. It was decided to wait until the morning to allow the ventricle to recover before it is insulted again. The next morning the ejection fraction had climbed to 35%. He had done so well that the surgeon wanted to wait one more day to be sure the kidneys had survived the insult. It seemed reasonable. Luck has much to do in caring for patients. For no particular reason the cardiologist went into the room and in the 3 minutes of staring at the monitor he noted a slight increase in heart rate and pulmonary pressures. The trend was real and the surgeon was asked to operate in an emergent manner. Just as the child was being wheeled to the operating room he arrested and was quickly resuscitated. He was crashed on the bypass machine and surgery completed. Now came the wait for the patient to wake up and see if the heart was irreversibly damaged. These are the times that one wishes for retirement or anything to get away from the sadness of defeat.

The child awoke and his heart completely recovered. Many of these children with anomalous coronaries simply die. Prayer along with a good team of doctors, nurses, saved this young man. It is truly one of God's miracles.

Every one needs rules to live by. Physicians are no different. Some of their rules may apply to the reader.

Chapter 13
Rules of the Houckster - How to Live With Your Heart

How to approach patients and life in general:

1) Never order a test if you are not prepared for the results.
2) When taking care of patients - never trust your mother - look at the test results yourself.
3) If symptoms persist, perform an angiogram regardless of how many tests are normal.
4) Before doing the above step, always perform one month trial of proton pump inhibitor and elevate the head of the bed 4 inches.
5) Consider patients to be a well-liked family member and treat them with love and respect.
6) Disagreeable patients and problem patients should be seen frequently (but not necessarily by you) spread them among nurses, dieticians, PA's, NPA's, social workers, priests and anyone you can find. It may be cheaper for the government to pay someone to be a friend to these individuals.
7) Be prepared - patients will do anything and everything to de-rail your treatment plan.
8) Always look at the old films before doing a new cath.
9) An 80 year-old female patient comments during a clinic visit, "Sex is not that important. It helps when times are difficult."
10) Fall in love daily. Just keep your hands to yourself when you are not at home.
11) Internship is a time to make mistakes and learn. Just don't make too many.
12) Smile when you are feeling depressed. Your brain will eventually think it is happy.
13) When you don't know what to do; stop what you are doing and pray.
14) It's good to be you. It is better to be the Houckster.
15) The enemy of good is better. (Does not apply to 14)
16) In taking care of patients be pessimistic and expect every bad outcome so you will be prepared to deal with the worst. When the worst does not happen you can be happy.
17) When on call and attempt to go somewhere with your family always take two cars. If you all go in one car, you will be called.
18) Bad admissions and bad times always come in threes.
19) Things always get better.
20) God has the final say.
21) Trust in the force
22) Everything should make sense and fit together. If it does not explain everything, keep thinking before acting.
23) When you decide to do an angiogram, imagine that the patient died during the procedure. Your decision is correct if you would have proceeded anyway. If you think of an alternative to the angiogram do the alternative first.
24) Hand-written instruction to patients – Every patient that you encounter should leave the office with a handwritten note that includes their lab, problems and what

you want them to do. It does not matter how smart you are if your patients do not follow your instructions.

25) Patients only remember the last thing they are told. Be sure it is the paper with their instructions.

26) Before every procedure empty your bladder.

27) My daughter Danelle's definition of a good citizen is a person who cares for everyone and holds no judgments. They are involved in their community, their church, and their school. They have flaws but are always striving to improve. They take leadership roles and have a positive attitude. A good citizen is not afraid of hard work that reaches a goal that benefits everyone.

28) Most importantly – keep a clean pair of underwear in your desk for those days when you are in the hospital all night with challenging patients.

Love

Love is like a grapefruit tree. When I first met my wife, she had recently planted a grapefruit seed from a coworker's leftover lunch. She had planted the seed to use as a science experiment in her classroom where she taught gnarly sixth-graders and prepubescent seventh graders. After many months of watering, the grapefruit seed germinated. It was an attractive houseplant with the deep lush green and spade appearing leaves. We soon married and the grapefruit became our joint possession. The grapefruit went to Chicago while I attended medical school and survived several attacks from our household cat named Tigger. This plant, along with Tigger, traveled nearly 1500 miles to San Antonio. Most of our other plants died due to the concentrating ability of the back car window on sunlight.

Southern Sun was much more acceptable to the grapefruit, which continued to grow and eventually developed small spikes on its trunk and branches. The tree became large enough that it was moved outside in a protected area near the house. The winters in San Antonio are quite mild but there would be several freezes during the winter months. At my wife's urging I would go out on the coldest day of the year to cover the plant. I would be rewarded with pinpricks from its thorns. The tree even survived a 13 inch snowfall a 500 year event. The tree continued to grow, and eventually it was time to move from San Antonio to Georgetown. The tree was painfully dug up and came with my wife, myself, 11 year old son, 4 year old daughter, and numerous cats. The tree was replanted in Georgetown. The first year the tree survived out in the middle of the yard and was covered with Christmas lights and bed sheets to prevent from freezing. One freeze however came quickly and caused the grapefruit to die to the ground. The tree had been approximately 5 feet tall when it came to an untimely death. To my surprise, the next spring sprouts came from the ground where the tree had been and it appeared to be a rebirth of the grapefruit. The grapefruit was again dug up and was placed next to the garage where could be better protected in times of weather extremes. I did not really like the tree as it became bigger and more difficult to protect from the elements, but to appease my darling wife I would risk life, limb, and pokes from thorns to cover the tree in adverse weather.

The grapefruit tree grew larger than the one story garage. This seedling had traveled with us through several moves, survived bone chilling freezes, and multiple re-plantings. I have been frost bit covering the tree with wraps of everything imaginable, Christmas lights, and pecan leaves. I have been scratched by sharp thorns and fallen off of ladders while attempting to rig a protective curtain. Each time my lovely wife would blink her twinkling eyes and I would respond to save that darn tree. After 30 years the tree bloomed. I pollinated the bloom myself by making sure the pollen of the stamen arrived on the pistil. There was only one grapefruit after 30 years of anticipation. The disappointment faded after I learned from a 90 year-old patient that this was typical of a fruit tree going through puberty. We all salivated and watched the fruit grow to tremendous size and the green change to shades of yellow. We decided that the third week in November would be our harvest.

My devious wife had a plan. She wanted to take the seeds and begin two new trees that would be given to our children for them to protect, move and be a living payback for the aggravation the children caused. The day before our harvest the lawn workers came and the grapefruit disappeared to our great sadness. My wife's plan to propagate the tree with appropriate pay back to our children had to be put on hold for another year.

The tree was amply covered for another winter season. Old blankets were dutifully hung on the tree. Multiple trips in ice storms, wind, and cold were required to keep most of the branches from becoming shriveled. The spring brought a bountiful supply of sweet pleasant white blossoms and numerous bees. Fertilizer spikes were driven into the ground by the drip line and the pool was backwashed into the tree to provide plenty of water. The green fruit increased in size and began to turn light yellow in late November. The fruit hung in clusters and that is the origin of the name grapefruit. The fruit grows in clusters and has a similar appearance to grapes clustered together on the vine. There were more than 100 fruit usually hanging close to long spiky thorns. The first freeze came early during thanksgiving week. My wife and I scurried as the cold front blew from the north to pick the fruit from low hanging branches. Some of the fruit were put into cold storage in the garage.

The grapefruit was known as a white seedless variety. Seedless seems incredible since it was grown from a seed. Seedless means there may be only one seed or no seed per fruit. It was then I learned that my wife did not like grapefruit and had only one segment of one fruit of the entire tree. Marriage is always filled with surprises. Apparently she enjoyed more my labor over the last 30 years and not the fruit of my labor. The 30 year venture to harvest was only appreciated by me. The fruit had a sour juicy taste. The amount of juice was incredible. The rind of the fruit was nearly an inch thick. I thought the rind could be used as an astringent.

I ate grapefruit for the next 4 months and enjoyed every juicy sweet tart fruit. The fruit eventually developed dark brown patches as they aged in the garage refrigerator. People were amazed that fruit was real and inquired as to what it represented. It represented love. The rare seeds were collected and placed into pots and watered daily. Grapefruit were given to friends and sent across the country to family with the request of propagation. It took a very long time until a small green shoot sprang from the ground. My children's fate was sealed and they would soon be presented with the responsibility of caring for a grapefruit of their own. My plan was to eventually have one of the offspring of these plants go to Mars. Love is like a grapefruit. It takes work, commitment, and includes disappointment and rewards.

The next winter, the tree was not covered because of its huge size. The freeze killed half of the tree, but it re-spouted. I had no hope of fruit and even after many months my spirits were again lifted when green spheres were again seen deep within the middle of the tree where even the winter freeze could not spoil a 30 year adventure. Just as in marriage, I learn how to be surprised. A new plan, perhaps temporary housing will be made to see if the tree can live another year and reward us with sweet juicy fruit.

Chapter 14
Death and Dying: A Natural Unspoken Event

The chilled family room adjacent to the emergency room was out of earshot from the multiple noises in the emergency room. There were 5 family members: a wife, two sons, a daughter in law, and a family friend. The chaplain was present and the mood was somber. Two hours ago the patriot of the family had been telling jokes at the table when he suddenly collapsed wedging him self between a chair and a bookcase. A silent shock followed by a horrified "Oh, No!" that ran simultaneously amidst the family who was visiting for a family birthday. No one knew CPR. After 30 seconds of panic and realization that this was not a practical joke, 911 was called. It was difficult to give directions. Panic tended to erase common facts from the brain. The dispatcher gave instruction to start CPR.

Emergency medical services arrived 15 minutes later and promptly began resuscitation efforts. The prognosis was not good because the patient had been down for 15 minutes and the initial ECG strip demonstrated asystole. With every minute of a cardiac arrest, the chance of survival falls. Survival is optimized by effective bystander CPR, EMS arrival within 10 minutes, and if the ECG strip shows ventricular tachycardia. If a single shock restores the patient, prognosis is good for survival; however, brain function may still be in jeopardy.

Unknowledgeable CPR can sometimes be effective. The general public shrinks from doing mouth-to-mouth resuscitation. New guidelines for resuscitation have now de-emphasized mouth-to-mouth resuscitation in favor of effective chest compressions. There is enough stored oxygen in the blood to last for several minutes if the blood can be moved to the oxygen hungry brain. Chest compressions are more important than breathing, especially in urban areas where EMS crews respond efficiently.

One individual out of desperation used a toilet plunger on the chest of his collapsed father. His father was successfully resuscitated and the communication of this success eventually resulted in a mechanical plunger machine. When an individual dies there is electrical uncoupling of the heart muscle. This can be restored with a simple shock. As this electrical uncoupling proceeds, it results in a hemodynamic collapse when the heart is not effectively filled and even if coupling is restored the pump cannot work effectively. Eventually, metabolic problems of acidosis, hypoxemia affects the individual cells resulting in the spirit leaving the body for a better place.

Atropine, epinephrine, intubation at the scene and vigorous CPR was successful in regaining a rhythm with minimal blood pressure. Fredric, a 58 year-old German immigrant was transported to the emergency room to a waiting code team where consciousness was never regained. Blood pressure could not be maintained and after one hour and 15 minutes of intense effort the code was called and the time noted for the death certificate.

The family was in shock and knew to expect bad news. The task of informing the family fell to the physician in charge of the code team. In this case, it was a third-year resident who was in charge of all codes for that evening. He had been trained for the last two years by observing his senior resident or staff physician break the news to families that their loved one had passed. There was no easy way to break the news. The only way to present the news was to be direct and understanding.

"I am Dr. Horrester, and are you Mrs. Fredrich Schwartz?"

"Yes." Mrs. Schwartz replied fearfully.

"And is this your family?"

"Yes," a quiet knowing yes.

"I am so sorry to inform you that we were unsuccessful in our efforts to revive your husband."

Silence, grief, tears, wails and family hugging ensues. Doctors and chaplains lay on hands to give comfort. Time seems to stop. The immediate grief reaction turns to painful silence.

"We are not certain of the cause of death but suspect he may have had a pulmonary embolus, a blood clot that traveled to his lungs and prevented the blood from flowing into his heart. His left leg was swollen and we suspect the cast he had been wearing for a fracture in that leg may have predisposed him to a clot. Do you have any questions or concerns?"

After a long silence, "No."

"If you think of anything in the future, we will be happy to answer your questions as best we can. ----pause--- I am sorry to trouble you with some other questions. We ask these questions every time someone dies. We would like to know if you would like an autopsy. The purpose of this is to help determine the cause of death and determine any other hidden medical condition that might help some other family members. It will give us better understanding as to what happened. We apologize for asking this of the family and you may have time to help make this decision. --- pause---- There will be some people who will help you with some paper work and arrangements. ---pause--- I am so sorry for your loss."

The family was quiet. It was time for the new leader of the family to take over and help with the decisions that need to be made. The chaplain was there to help and sometimes no one would be willing to step to the plate. The doctors left.

What a happy death for Mr. Schwartz. Family surrounded him during a happy occasion, while he was laughing at his own joke. One could imagine him looking down from that bright white light watching his family's amused faces as he hit the floor. He did not suffer for an instant. His death was tragic because it came too soon. Except for it being to soon, it is a death we all would like to have. No foreknowledge, no pain, or suffering. We are not prepared to be born, and we are not prepared to die.

I do not like death. I am graded by mortality and it is published on the web. Death is my report card. I do not like funerals. I do not like telling family members that I am sorry. I may be guilty of hanging on too long. I am reluctant to pull the plug if I sense there is even a tiny hope of survival. The sicker the patient - the greater the challenge. I do have my limits, and the limit is brain function.

Families and patients need to have guidance in making decisions on critically ill family members. The average family member is unable to filter through the large amount of information that they are given to make accurate decisions. They have to depend on their doctors giving them a clear understanding of the patient's prognosis. Many times, family members have other emotions that prevent them from making decisions. Usually, the most adamant family member is the one who is located the farthest from home and had the least participation in their parents care. This inability to participate in the care is often associated with guilt.

In an ideal situation, families can rely on a living will to help surviving members sort through some of these difficult issues. It is better to have frank discussions with family members as to what a patient wishes. The prospects of death, infirmity, are not pleasant and therefore rarely occur. The family may be unaware of the living will and once the patient becomes unable to make their own decisions the family then becomes the decision maker. In some states, emergency medical personnel are required to ask if a living will is present before they resuscitate the patient. This seems a bit unreasonable since the act of calling EMS is a volunteered withdrawal of the living will.

Rosie was an 83 year-old female who had chest pain for 12 hours before calling her daughter. She developed more shortness of breath and when her daughter arrived at the home she was found on the floor. EMS arrived within 20 minutes after being called by the daughter and began resuscitation efforts. The initial electrocardiographic tracing demonstrated ventricular fibrillation. Survival after ventricular fibrillation cardiac arrest drops 10% per minute. After one hour of on the scene resuscitation, a blood pressure of 70/80 was obtained. The patient was brought to the emergency room where she was further stabilized. Pupils were dilated and non-reactive to light indicating the brain was no longer functioning. This physical exam finding, however, is no longer reliable after the administration of Atropine since Atropine also dilates the pupils. Further history revealed that she experienced progressive chest pains over the last month and had pain for greater than 12 hours before asking for help. It appeared that she had been down for at least 15 to possibly 30 minutes. Her resuscitation was lengthy. She had diabetes mellitus and mild renal insufficiency and a previous stroke. Her physical examination revealed an unresponsive elderly female who was mildly obese. She demonstrated no purposeful activity when stimulated by painful stimuli such as placing an intravenous access. This movement indicated no white matter function of the brain.

Physical exam findings during the first 24 hours after a cardiac arrest are not reliable due to brain swelling from the hypotensive injury. Patients need to be observed for 24 hours before a prognosis can be estimated. The longer it takes for the patient to wake up and make purposeful movements, the worse the prognosis. 24 hours of monitoring is considered necessary even in patients who had a poor prognostic resuscitation such as Rosie's. The longer it takes a patient to wake up the greater their mental deficits. After 24 hours of further observation, Rosie's condition had not changed–other than a few spontaneous respirations. The family was informed of her poor prognosis and it was recommended that she be made a "do not resuscitate" at this time because further efforts to maintain her life were futile. The family was uncomfortable with this decision and requested that further time be giving to allow for other family members to come to the bedside. The family's wishes were honored and two additional days of observation went by. There were multiple family meetings to discuss discontinuation of care. There was disagreement among family members and as a result there was reluctance to withdrawal of care. The futility of the situation was explained in a compassionate manner but fell on unrealistic ears. After 5 days the tube was removed and Rosie passed 15 minutes later.

The expense of dying in this country has risen. More money will be spent on nursing home care in 2030 than was spent on Social Security in the year 2000. Institutional care costs $75,000 per year. This is more than most patients ever made in one year. It is estimated that between 25% and 45% of all patients over 85 will need to receive this expensive care. This does not include catastrophic events like a myocardial infarction or stroke where the hospitalization cost could approach $100,000.

In contrast, in the adjacent room was another near death event. Melinda was 70 and was undergoing cataract surgery when she had a ventricular fibrillation cardiac arrest. She was resuscitated within 5 minutes but did not awaken as expected. She was intubated and taken to the coronary care unit for further monitoring. She had no known cardiovascular disease but did have a family history of sudden cardiac death. She was the oldest survivor in her family. She had been given erythromycin for bronchitis. She had no other co-morbid conditions. Melinda did not respond but demonstrated no posturing, and had spontaneous respiration. Posturing is an involuntary reflex that help determine the level of brain functioning. Decorticate posturing is flexion at the elbows. Decerebrate posturing is extension at the elbows and has a worse prognosis. Her ECG demonstrated a long QT which was the reason for primary arrhythmia instead of a myocardial infarction. No injury current was present indicating no cardiac damage. The erythromycin, a benign common drug can be deadly in patients with a prolonged QT. She apparently did have a living will kept safely at home. A daughter arrived into the unit distraught and demanding that her mother never wanted to suffer this way and she demanded the ventilator be discontinued.

It is difficult to communicate to hysterical family members. Education of the situation is necessary to help defuse the conflict. Some ears may hear but the brain is incapable of processing new data. An attempt at an explanation and education proceeded.

"We are sorry for your trouble and appreciate your desire to fulfill your mother's wishes. It is not time for that decision and we will help you in letting you know when that time comes. We are very concerned that she has not regained consciousness. We have to give her time to awake and she has a good chance of having a full recovery. She did not have any heart damage and her resuscitation was quickly and efficiently performed. The quicker she awakens, the better her prognosis. She requires at least 24 hours and sometimes longer since patients do not always behave like text books."

"I don't care --sob—She would not want to live like this. See-she is choking on the tube and looks like she is in pain."

"I can reassure you that she is not in pain and will have no recollection of these events. Unfortunately, We have been under these terrible conditions in the past. We do not know how this is going to turn out but we need time to give her a chance to recover. I could not sleep tonight if we took out this tube. It would be the wrong thing to do. She needs time to recover. I know this is very hard for you to see your mom in this condition. We will help you in any way we can, but it is too early."

The daughter was not listening. She exhibited silent anger and grief, turned abruptly, and left the room.

"Excuse me, but sometime in the near future we suggest you and your children have electrocardiograms to see if there is a family predisposition to ventricular fibrillation. We will talk again when you are ready."

The next morning, Melinda was awake and the tube was removed. She had no deficits and after a defibrillator she went back to her home living independently. Her family was screened and two other defibrillators were placed.

More of my patients are asking a new question for different reasons. They are asking how long they are going to live. They are asking because they are living longer than they had expected. Savings and fixed income is causing tension among the retired on fixed incomes. I tell some of them to go and get a job with a mischievous smile. The answer deserves better thought than a mischievous smile. As you get older some expenses get less like mortgages and taxes. Vacations are not taken. The unexpected looms greater. Air conditioners fail, water heaters need to be replaced, and household upkeep never stops. Chores that people liked to do no longer can be accomplished because of injury and weakness. These chores are then contracted to outside help. Long-term care insurance is a necessity. A nursing home can eat into savings at a tremendous rate. Home help is also very expensive. Eighty year-old patients never planned for these nasty occurrences. They plan for the normal expenses but have never planned for prolonged recovery after a stroke. The return of parents moving into children's homes is happening at a greater rate. The economics of aging is too scary to study. The baby boomers will soon have to contend with all of these new life stresses. Failing parents and bankrupt children will redefine the extended family.

Determining end of life is a very difficult task. No one would have predicted Fredric Schwartz would have died. There are insurance actuaries that can give an estimate of longevity based on age, smoking and risk factors. They are accurate for populations but fail for individuals. End of life is particularly difficult in congestive heart failure. The published mortality of 50% in five years has been reduced. The elderly ninety year-old with heart failure died within weeks. Now they seem to go on forever. Prediction about death is very difficult. Heart failure patients that are sent to hospice are felt to have less than 6 months to live. In twenty-five years of practice, I am unable to make an accurate prediction about dying.

Kidney function is predictive and a bun >43, creatinine >2.75, a Blood Pressure <115 is predictive of a 1in 5 chance of dying from heart failure during a hospitalization for heart failure. Even this mortality has changed since the observation was first met. Ejection fraction, which is the squeezing function of the heart, can predict prognosis but is not reliable for predicting death. Heart size is a better predictor than ejection fraction. Heart failure patients are YoYo's rising and falling many times before death becomes victorious.

In 1993 Manuel was 76 years old and was first referred to the Houckster for congestive heart failure. He was having dyspnea with exertion. He worked at a supply parts store and was constantly on his feet. He had increased cough and because of persistent symptoms eventually came to his primary physician. He had crackles at the bases and was initially started on Lasix to relieve the symptoms. During evaluation with the Houckster he was noted to have an ejection fraction of 20% and moderate to severe aortic insufficiency and a dilated aortic root. He also had a 4 cm abdominal aortic aneurysm. Valve replacement surgery with an ejection fraction of 20% secondary to aortic insufficiency was not considered an option because of the poor left ventricular function. Manuel never had any chest discomforts and coronary angiography demonstrated no significant obstructive lesions. He was placed on medical therapy and his heart failure medications were titrated. Initially he felt poorly on the medications and his medicines had to be decreased on two occasions before they were again increased. Eventually he was on full beta blockade with Carvedilol 25 mg two times a day, Lisinopril 20 mg two times a day and Spironolactone 25 mg a day. The patient came to a clinic and had no complaints and ejection fraction at this time demonstrated a near normalization of his left ventricular function. His ejection fraction was now 50%.

The dilemma that now faced the Houckster was: Should this patient be sent to aortic valve and root replacement since he was now a surgical candidate due to the improved heart function? Since the patient had no symptoms it was decided to continue to follow him medically. Five years later, at the age of 81, Manuel is having fatigue and weak spells with blood pressures that were running typically 85 to 90 systolic. For this reason his medications were reduced. He then suffered a sudden cardiac death and was resuscitated successfully through rapid response of EMS. He underwent defibrillator placement and again had his medications titrated with resolution of the symptoms. With additional beta blockade Manuel became pacemaker dependent and his ejection fraction again fell to 20% with marked cardiac dyssynchrony. He was submitted for an upgrade in his defibrillator to a biventricular defibrillator. During the testing of the defibrillator he had electromechanical dissociation requiring pressure agents and again suffered mild brain injury. Very sick hearts may just get tired and stop pumping. A defibrillator can not help this condition and is the cause of death in many heart failure patients who have defibrillators. The defibrillator restores the rhythm, but the heart muscle does not respond to the organized electrical activity. Other than memory loss he survived his second episode of death. For the next three years he continued to work part-time at the parts store but became very symptomatic with fatigue, lightheadedness and weakness. He was again admitted to the hospital and had a blood pressure of 65 systolic. Efforts at improving his blood pressure were unsuccessful. His renal function had worsened to a creatinine of 3.0 and a BUN of 90. His heart was large, kidneys were failing, and he could not support blood pressure. Hospice was notified and accepted the patient for comfort care. Manuel had his defibrillator turned off before leaving the hospital.

Six months had gone by and Manuel again returned to the Houckster's office. He had been discharged from hospice because he had not died and in fact was doing quite well. His blood pressure was 120/80 and he had no symptoms to cause him any degree of discomfort. The Houckster examined the patient and felt that his heart function had improved with smaller chamber dimensions. Blood tests demonstrated his kidney function had improved to normal. The defibrillator was reactivated and his medications were again titrated. Manuel is now 87 and still without symptoms. His death was predicted three times and three times the prediction was wrong. He demonstrates that some patients with heart failure can behave like YoYo's bouncing up and down in their illness. The reason for his hypotension during his last hospitalization is unknown. Perhaps the intense neuroendocrine blockade by his medications caused a stealthy transient adrenal insufficiency or other neuroendocrine insufficiencies. I have even considered testosterone failure. His bowel may have been edematous allowing endotoxins to leak into the blood stream causing vasodilatation. We do not have a great understanding of these processes in the very late stages of heart failure. It may be due to promolecules failure to be cleaved into active peptides that may explain this hypotension, anemia of heart failure, and the elevation of BNP.

Five years ago, it seemed that heart failure was easy. Seventy percent of patients responded and stabilized and 30% went onto die. The YoYo affect was not recognized. By upgrading devices and personalizing therapy, an occasional radical surgery could restore even the patients with the worst prognosis. Treating anemia, using testosterone, Viagra, and treating patients with IV infusions seems to pull them back from the brink. Artificial hearts that allow the heart to rest and repair is slowly leaving the newspapers and becoming main stream.

William was an engineer and was always driven. Although he retired 3 years ago at the age of 73, William continued to have an active role in his consultation business. He had imagination, purpose, and a bad heart. William had a previous heart attack and bypass surgery. Although, he had never passed out, ventricular tachycardia was documented on a Holter monitor. William was skeptical of his health care providers because they would tell him how sick he was but he did not feel bad. When patients feel well they are reluctant to take medication. Recent onset of shortness of breath that kept him from his normal activities changed his mind. Visiting the Houckster for more than 12 years eventually resulted in trust after initially being skeptical of the recently retired Air Force doctor. The idea that his doctor had been an engineer before going into medicine was appealing. Slowly with education he began to take medications and he improved. A defibrillator was eventually implanted. The third lead, which we were just beginning to implant, could not be placed in the lateral wall since there was not an acceptable lead position. The patient received an RV apical lead and the third lead placed into the left ventricular outflow tract. His ejection fraction improved. One angioplasty was required. At age 85 he again had a decline in cardiac function. The decline was realized by a decline of ejection fraction to 20%, fatigue, and renal insufficiency with a creatinine climbing to 3.7 from a normal value of 1.0. William had been building an airplane, but had lost interest in completing the project.

The Houckster offered him an upgrade of his device to a true lateral lead. This would require the surgeons to place the lead through the lateral chest wall. He was given no guarantee. He knew he was declining and agreed to the procedure. With the third lead properly placed in the lateral wall his ejection fraction improved from 20% to 35% and his kidney function improved. He continued work on his plane and was frustrated that he did not have a valid flying license. A suspicious gleam in his William's eye, a determined chin worried the Houckster. The home built plane was not just an aircraft- but a racer. The plane could easily out fly him.

When do you call it quits in pulling tricks out of your hat? Age is not a good judge. Attitude may be negative when you are feeling sick. Attitude may also improve with an improved body. Dementia is the final stumbling block in an attempt to raise the YO YO one more time. I often wonder if dementia is not a manifestation of illness, losing the zest in life. The brain has stem cells and when these stem cells become depleted life drains away into oblivion.

Predicting death in heart failure is difficult. Discussion of dying is as necessary as lectures on salt. Defibrillators are soul catchers. Ventricular fibrillation is painless release of the soul. We have invented special pacemakers that can detect ventricular fibrillation and deliver a quick shock that will restore life and catch the soul from escaping.

Patients who have ejection fraction ≤ 35 % qualify for a defibrillator. The only contraindication is if the life expectancy is short due to cancer, or heart failure, or significant dementia. The discussion over a defibrillator can go as follows.

"I want you to think about something before our next visit. You qualify for a defibrillator which is a special pacemaker that will shock you back to life from a death causing arrhythmia. Some people call it a soul catcher. It depends on your point of view. If you are willing to meet your maker when he calls you home you may not want this device. In turn, if you believe that it could help and you want to increase longevity, you may want to have it implanted. There is no right or wrong answer and it depends on your preference."

Before the issue of death and dying is another potential stressful situation. This stressful situation is best known as a child trying to get their parents into a nursing home. The ultimate of life stress is going to the nursing home. You have to give up all of your hard earned stuff. You have lost your own autonomy and privacy. Retirement communities are springing out of the ground in the south. Most of them should have large assisted living quarters and long term care facilities, but they don't. It reminds the residents of what is coming. Because this is an unpleasant thought they are absent. The time to move into one of these nursing types of homes is when safety demands. If you fall down and cannot get on your own feet by yourself it is time to go. If you can't remember that you took your medicine it is time to go. If you are aggravating your neighbors and they are not getting sleep watching your house, it is time to go.

In Summary:
1) You can never be prepared for death. An expected death is still tragic.
2) Death is my enemy.
3) Death can be my friend when suffering is great.
4) A plan for your death should be openly discussed with your immediate family or executor so everyone is clear on your wishes so an out of town relative does not interfere with your last moments on earth.
5) Defibrillator discussions should be direct to the issue of patient preference in choosing mode of death. If heart failure cannot be improved and you are suffering then death from ventricular fibrillation may be welcomed.
6) Physicians should always discuss patient preference if their heart should stop, or they can no longer breathe without mechanical assistance. The time for this discussion is when the patient is feeling well and not in an emergency.
7) A living will should not be honored during an admission for myocardial infarction if there is no other terminal illness. This is not a terminal illness and can respond to simple painless procedures. An electric shock should be delivered immediately to restore life.
8) If you do not want to be resuscitated and intend on meeting your Maker do not come to the hospital.
9) If you fall down and can not get up a nursing home should be considered.
10) Communication with your doctor, friends, and family should include death, nursing homes, and unpleasant unfinished business. There should be a holiday assigned each year (**National Death Day**) for the specific purpose of communication. A nice meal followed by frank discussions with all in attendance from 2 year olds to 98 year old family members. Anything can be stated and discussed. After the discussion a keg should be ceremoniously tapped to let all the fears of the future flow away from the family now that the unspeakable has been spoken.

Chapter 15
Murray - mystery solved or initiated

The Negritos, a 35-year-old pygmy and his brother, were perched 25 feet above the jungle floor watching the camp. Below, he could see the campfire with the four Americans sitting in a semicircle. Beyond the camp little licks of light illuminated the jungle. The moon shaped like a bowl was keeping away the rains. Below the moon, high mountains leading into Baguio city were outlined. The Americans had been walking for 6 hours before they stopped to camp. The great Lion Head was 20 miles behind. The nearest dirt road was now 10 miles away. The Negrito had not slept for 18 hours. The Sly Father had not paid him to sleep.

In the camp, Murray, the Houckster, Collette, and Tracey finished their MRE's (meals ready to eat) and were relaxing before going to sleep. Murray was quietly staring into the fire. Collette and Tracy had tired silliness and were trading boyfriend stupid trick stories. The Houckster was already snoring in an effort to keep the wild animals away. Collette could not believe the transformation of Murray. When she had first met Murray, he was a broken old man. He had a big heart and that was not in the sense of giving. His recovery of heart function had been miraculous. She was getting used to miracles. Miracles should never be taken for granted.

Murray continued to take his medications and stopped smoking. He had received a bi-ventricular AICD, which had restored cardiac synchrony and completely resolved his mitral regurgitation. During the course of treatment, his atrial fibrillation converted to sinus rhythm. This in itself would have been considered a miracle ten years ago. The real miracle was Murray's renewed zest for life and purpose. An old man can shrink from life giving into the pain of aged joints and forgetfulness. Murray would not submit to the typical picture of an aging old soldier. An exercise program was initiated early in his heart failure treatment. He bought a stationary bicycle and was perplexed when he could only manage 5 minutes. The Houckster told him to reduce the resistance and try to add a minute to his riding duration every several days. He was told to attempt the exercise multiple times a day for a total of one hour per day.

The old soldier, predecessor of the Green Beret's, increased his duration to 30 minutes two times per day and began to feel life drain back into his soul. Murray noticed that he was a bit wobbly when he got off the bicycle and decided to walk for 30 minutes. Within 4 months he also started a free weight program. Great strives were made with his heart and he was no longer limited in his breathing. He began to train in earnest. Murray lost 30 pounds and converted another 30 pounds to lean muscle. Swimming was added to his daily routine and this helped him regain flexibility. His muscles and joints responded to the great number of stem cells his body was reproducing. As he rejuvenated his heart, his mind and body became younger. Murray was a driven old man. He found renewed purpose and wanted at least one last adventure.

His purpose and quest was to give an answer to a recurrent dream. Nightmares about the Philippines and some of the terrible acts he had committed to stay alive would interrupt his sleep. The norm was to awaken in a pool of sweat. The darkness of his room would be confusing and initially he thought he was back in the dark cleft hiding from his Japanese pursuers. His dreams and reality became blurred over the last 6 decades. Murray knew there was something important in that cleft but was having trouble recalling fact from fiction. The fever he had may have lead to hallucinations, but he was drawn to the past with a feeling that something important was in that cleft. The only way to resolve this issue was to return to the narrow passage and see for himself.

The Houckster, Tracie, and Colette were the catalyst. They had saved his life. As Murray shrank from the broken man with heart failure and became a man of respect, his rambling wild stories became an impressive awe. Tracie, the skeptic, researched his activities and determined he was true to life. She had great respect for the old gentleman. It took a lot of convincing to take this adventure. In the final analysis, the safari through the jungle and mountains of the Philippines had to be better than a cruise or a skiing adventure and she wanted something to impress her children.

The Houckster had a different reason for coming. He had not wanted to go to the Philippines on his first assignment. The assignment was the lowest in Air Force cardiology circles. Coming out of training he was an easy prey for the assignment office. While in the Philippines he found great joy in its people. The volcanic islands and many cultures were fascinating. A trip back to see this troubled paradise was really irresistible. He wanted to see his old residents and again smell the market place. The smells of a third world market ranged from pleasant to rank. Un-refrigerated fish and meat, in addition to the sweet smells of baking and flowers mingled with the scent of crowded people and pets gave a distinctive aroma. He had hoped things improved for the people. While stationed at Clark Air Base, he had not seen a great deal of social progress from World War II. Terrorism, bullies and corruption kept the people from real freedom. The schools were a saving grace.

He often thought about his third world life when he would look at the good luck bonnet he had purchased. It was a fertility and good luck headdress that was composed of a monkey head skull, boar tusks, feathers, beads, and a small statue that represented fertility for crops and children. It had a colorful woven red band. It was expensive - costing 15 dollars. He hadn't the heart to haggle. It was with this hat that he found out just how superstitious Residents in training can be. Wilford Hall Medical Center was a busy hospital and received patients via Air Evacuation from all over the world. It was a great place to train but was not very conducive to sleeping through the night. It would not be unusual for cardiology to receive twenty admissions in an evening.

The Houckster's team had gotten hit hard for three call nights in a row and this was not allowing any time for teaching. The Houckster pulled out the headdress and put it in front of a fan so the feathers would blow across the monkey head. He told the residents they would have good luck since the moving feathers would scare away the evil spirits. That night there were only three admissions. The next call night the fan was turned on and there were no admissions. The Houckster had enough. The residents were getting lazy and they had not learned anything. He locked the doors and made sure the fan was off. Thirty-five admissions a new record came from the emergency room and a total of 4 Air EVAC 727's. The power of the headdress was now proven beyond a doubt. The residents were hooked on the hat. The superstition of the hat is similar to copper bracelets for arthritis. If you are feeling good while you wear the bracelet and feel bad when it is missing, you will believe in the power of copper. It is another placebo affect. At the end of the year, the Houckster discovered his residents would sneak into his office and turn the fan on the feathers late at night and turn the fan off in the early hours of the morning. They continued this practice for the rest of the year. They had great respect for the hat.

The hat was later used by a number of young couples that were having trouble conceiving. They would come to the Houckster who would give them instruction in the use of the headdress. It was important to take the females temperature. It was more important to abstain from sexual activity for 5 days before an anticipated temperature rise. At the time of temperature rise, the female was to wear the hat during relations. Positioning is important but not for the purpose of this story. The hat was successful every time although it did require invitro fertilization on one occasion. This one extreme measure was thought to be due to non-compliance with abstinence. The hat still sits on a shelf in the Houckster's office near a fan next to a ceramic jar labeled "ashes of problem patients."

Colette was in the Philippines because of a failed relationship. Female doctors have imposing schedules and boyfriends have to be particularly thoughtful and caring. Colette just wanted to get away and be with people she enjoyed. She was ready for any emergency. After all, she was training to be an Emergency Room Physician. She had liked Murray from the beginning and was kept up to date on his progress by Tracie. She was uncertain of his tale but was ready for an outdoor adventure. It seemed paradoxical that someone with so much beauty would like to be in a position where there were no showers for days on end.

Murray stirred the fire sending embers to the jungle canopy above. After the girls fell silent, Murray asked them if they knew the significance of the date 21 December 2012. Collette volunteered it was her birthday and she would be 22. Murray's grey green eyes and half smile made Colette giggle, he had seen through her obvious lie but was too kind to comment. He looked up at the Milky Way and said that the Earth's alignment and precession, the Sun, and the center of the Galaxy would be aligned. Collette sighed and said she had trouble with astronomy. This date was to be the end of the world as predicted by the Mayan civilization. They were great astronomers and had a calendar that is more accurate than the modern calendar. They were able to predict eclipses and the motion of Venus across the night sky. They did not look at time as beginning and end but as a cycle just as the moon circles the earth and the earth circles the Sun and the Sun circles the Milky Way. The date 12/21/2012 was predicted as the end of a great cycle more than 5000 years ago.

We would know more but Spanish Priests destroyed the Mayan library. Through visits to libraries and examination of remaining documents the old soldier was able to examine some of the few remaining paintings that depicted the calendars and predictions. If we find my lost cave I believe there will be strong similarities to the Mayan culture. The Mayan's advanced knowledge of astronomy was precise and difficult to determine from Earth based observations. Some fringe scholars feel the only way the calendar and predictions could have been made was from observations from space. The Nazca desert in Peru has ancient geoglyphs that were not discovered until airplanes in the 1920's flew over the site and revealed monkeys, hummingbirds drawn in gigantic scale on the ground. How did the native people make these pictures that can only be seen from the air? A more perplexing question is: Why they made these figures? Were they attempting to communicate with deities in the sky or were they acknowledging a people who came from the sky?

The figures were designed more than 2,200 years ago. The designs are made with simple tools and have been preserved in the desert for two millennia. The motivation for these people to make these huge designs that covered over 50 miles with the largest figure being 900 feet long must've been tremendous. It is somewhat inconceivable to me that these people had free time to spend making these works of art that they themselves could not even see. The battle for survival would seem to preclude free time. Like many other ancient civilizations, the Mayan's experienced cycles of renewal and decay. Detailed information about societies has only existed since the development of the printing press, however there may be many ancient secrets, which have to be rediscovered in the modern era.

Murray further commented, "In my case, in the cleft that I sought refuge, I believe I saw some symbols that are similar to the Mayan calendar and a Nazca lines. The possibility that my cave and ancient secrets of the Mayan culture and the Nazca people could be related is very improbable. These cultures could not have traveled clear across the globe in ancient times."

Tracy piped up from the campfire and commented "Murray, are you trying to tell us that we are out here chasing aliens?"

Fire seemed to glow off of Murray's beard stubble as he looked up into the sky and simply closed his eyes. His last comment before retiring for the evening was that they should all get a good night's rest since the trek would become more difficult and potentially dangerous. They were entering the homeland of headhunters. Both Tracy and Colette snickered softly as they went off into their tent hoping they could get some sleep above the snores of the Houckster.

The next morning, Camp was broken with extinguishing of the fire and policing of the area an attempt to make it look untouched by the four travelers. Before leaving, Murray made a bird call in the direction of the jungle canopy. Climbing down from the treetops Peti and his brothers startled everyone except for Murray. Colette and Tracy instinctively screamed silently and moved closer to Murray. They simultaneously whispered into his ear, "Are these the headhunters? We thought you were kidding trying to give us bad dreams."

Murray chuckled and informed the duo that Peti and his brother had been their security guard for the last three days. He hired them to protect their camp from thieves. He had encountered their parents many years ago when he was in need of assistance. Their families had readily helped Murray perform dirty tricks to the occupying Japanese. Murray was impressed with their culture and ferocity as a people. Murray helped their families by putting out of commission a band of Japanese who threatened their tribe. He also recognized that as a people they were kind, generous and happy. The headhunters of which he had spoken were no myth. These people were also a strong culture that had a dangerous flaw. When a young man wished to take a wife, he needed to present to their family a human jaw demonstrating that he can protect his new wife and family from future enemies. This is how the tribe became known as Headhunters. As one can imagine neighboring tribes would be in constant skirmishes. Tribes on both sides of the conflict had fewer males as a result of this practice and wives increased in numbers. Most of the work of survival fell to the women of the village. He thought both Colette and Tracie would be able to relate to them well. His jaw and Dr. Houck's jaw may appear to be a good trophy.

Murray learned of this culture decades ago and was quite surprised to learn the practice of headhunting still occurred (but with much less frequency). Headhunting generally occurred only during times of stress from drought. The Negritos had learned to stay away from the tribe's borders. Peti and his brother would scout ahead to determine if the tribes were looking for new trophies to expand their families. They would be paid extra for the next three days as they traveled through this territory. Every attempt to avoid contacts with the natives would be attempted. Their path, however, was through the headhunters' jungle. Stealth was not in favor of the novice explorers.

Peti and his brother disappeared silently into the jungle followed by a noisy Murray tribe. No roads equated to no economic development. If the Amazon is to be saved, no more roads should be built into the rain forest. Roads bring commerce and with commerce come clearing of the land and a change in the environment to suit the human occupants. Disease is reduced by sewage systems and development of purified drinking water to remove the nasty microbes. No roads meant life continues without change. The twenty-first century had no impact on native peoples without roads. The trip through the jungle now transformed from adventure to painful work. Encounters with snakes, wild boars, and a multitude of insects made everyone very cognizant of where their feet were placed. Clearing a brush to make headway was necessary in certain thick groves.

The Houckster inquired of Murray how he was navigating using no map and only a compass to guide his progress. Murray replied they were traveling between two mountain ridges that join together 6 miles from the present position. As long as he kept one mountain ridge on his left shoulder and one on the right shoulder he knew they were traveling in the right direction. The mountain ridges would eventually meet and they would be 2 miles to their destination when they began to travel uphill. The uphill trek would not be easy because it was quite steep. The ridge that they were attempting to locate would require 50 feet of free hand rock wall climbing for the first individual up the slope. Ropes can then be lowered to allow the rest of the group to more easily traverse the shear rock wall.

Sensing their fatigue, Murray decided to stop their trek for the day and set about making a camp. Three individuals would sleep in a tent while one would keep watch. There would be no light that may attract curious native people. It was reassuring that the treetops would be manned by the Negritos. Sounds of the jungle included chirping cicadas, croaking frogs and the incessant buzz of mosquitoes. Their clothing became stiff from perspiration with stains that began to coalesce into a scaly pattern.

Despite their sore feet and fatigue, Collette and Tracy were full of enthusiasm. The Houckster was beginning to doubt his judgment in coming on this trip. He was familiar with treks having accompanied his son and Boy Scout troop to Philmont and had gone through survival school in the Air Force. He was not as tough as he had imagined and thought of creature comforts at home. He enjoyed his time at watch, observing life in the jungle. When it was his time to sleep he tried to sleep on his belly to minimize his snoring and sleep apnea. Imagining the 40° winters and jumping into the heated pool swimming laps. Sleep would eventually come. No one heard the skirmish that occurred that night. The evidence for the skirmish was two pair of ears attached to a string dangling from a branch.

The trek became quite somber the next morning. Human life may have been given up to the jungle. Two young headhunters were tracking them and had been planning their attack. With silent blow darts Petri and his brother had felled the pair. They removed their ears and removed the poisoned darts. They were then at the mercy of the Jungle. The darts had not taken their life but had immobilized them. They still had shallow respirations. The pygmies had not taken their life but had reduced their chance for survival. The willingness of jungle would determine their fate. The amount of poison on the dart that was absorbed was not a very good predictor. They may awaken with no ears and be unaware of how this came to be. The blood scent may make them a victim of the Jungle. Murray hosted a town meeting to let everyone express his or her feeling and decide if the trek was to continue. No one wanted to turn back and run like children.

The next two days were without encounters and the expectations of reaching their goal were high. Colette was to be the mountain climber since she had the most experience. It helped to be small with good muscle strength. She had previous experience with free hand climbing and had great respect for falling. If you had met her in Sunday School you would never have imagined her to be a climber. She was able to find natural handholds and foot rests where others could not. She loved physical activity and spent most of her free time shared between studies and physical activity. She was ready and excited as the terrain began to increase in altitude. Looking back from where they had traveled, it did not seem like a great distance. The giant rock formation of a lion head could still be seen from their vantage and it seemed to be looking straight at them. The lion did not seem so ferocious. It seemed to have a regal acknowledgement, a proud father type of look. Camp would be made with the ascent planned in the morning. Spirits were again high and stories flew around the camp.

Murray was uncharacteristically verbal. His whole life was one of action and not of speech. He now told countless unbelievable stories of his youth. With Murray it was hard to tell fact from fiction. Embellishment would make the story better but the truth then became harder to believe. He then began to relate feelings and told of how he grew old and sick. It's hell to get old. We all look forward to retirement. Once we get there it is not what you thought it was going to be. It felt good not setting the alarm clock and knowing you could rise anytime you wanted. The reality is I could not sleep past six A.M. Then I had trouble sleeping, waking up every three hours to go urinate, micturate according to you doctors. The morning aches and pains with joint stiffness and back pain. I gave into those pains and stopped doing things thinking it was natural due to my age. When I lost my wife, my routine was forever changed. She was my companion, my caregiver, and my friend. I watched my friends disappear having their heart attacks, cancer and just getting lost from living. Many of them were already dead and living in purgatory without even knowing. Their attitude became one of existing and why they couldn't just pass on to the next life to be with their dead husband or wife. Some of them lived for their extended families. Some were heartbroken because of son and daughter's broken lives. Once you are a parent you are always a parent. There is no educational course in aging you just have to experience the act. No one wants to lose their autonomy and become dependent on others. We old folks were never taught to age and we do not know how to act.

Murray continued, "When I lost my wife of 45 years, I seemed to become a turtle living in a shell going at infinitely slow speed. As I got sicker I became more despondent and wanted to die. It is hard to die and getting sick is not a pleasant adventure. The medicines and medical devices put me back together. The real change was the understanding that if I wanted to feel good it was my responsibility to change. Exercise improved my body but cured my brain and outlook. I found I still hurt; but I did not hurt any worse whether I was active or not. Gradually as the training continued I began to hurt less and feel good. I would not give into the pain. On days when my schedule prevented me from doing activity I felt worse. It may not be the fountain of youth but exercise made me younger in body and mind. I feel stronger now then when I was meandering around this jungle looking to survive and cause mischief. I relearned that to live you have to have goals. The goal could be to find a lost mystery. The goal could be as simple as watching and caring for a garden, watching the plants spring forward and give new life. It could be the noon lunch at the nursing home spreading stories to other residents. If you do not have purpose you are just taking up space. I don't want to take up space and the fear of this happening was my motivation to improve."

After a brief silence, Tracie began to speak, hesitated, and then shrank. Murray looked at her perplexed face and had knowledge of her dilemma and replied. "Tracie, you are wondering if Collette should climb tomorrow based on a 55 year old hallucination." He paused, and then added, "I have been thinking about that myself."

Collette chimed, "It is my decision to climb and I feel the climb is reason enough for me to get to the top. I will be careful. If I find nothing the climb would be reward enough."

The conversation trailed off and the foursome went to their tents to get some sleep before the morning climb.

The next morning, the Houckster was boiling water for coffee as Tracie and Colette stumbled out of their tents. Looking puzzled at the Houckster, the young duo wondered about Murray. The Houckster smiled and slowly looked up the precipice. At forty-five feet up Murray was 10 feet from his goal. The ledge could not be seen from the ground. He had to make handholds for his final ascent. Murray wondered how he was able to scurry up this precipice when he was twenty. Being pursued by soldiers with guns certainly could raise adrenaline. He told Colette he did not want to steal her thunder. He had woken up early and decided to give the first ten feet a try. It was so much fun that he just kept going. He would drop a line so she could help him with the final climb. Colette drank her coffee and effortlessly rose to Murray's level. Once at Murray's level she could see the edge of the ledge 10 feet above.

Colette swung on the rope to the left and then to the right. She spotted a tiny crack that could be used as a handhold. Climbing was easy with the repelling stakes easily fitting into the crack. She reached the top and lowered another rope so Murray could go to the top. He saw the cleft in the mountain. He wanted to rush into the mountain but stopped. He turned and yelled to Tracie and the Houckster that they did not need to take all day. The dishes would still be there when they got back down. With a great deal more effort, Tracie and the Houckster made their way to the top. They had carried lanterns for the next phase of their adventure. The four were on the ledge and Murray pointed to chipped blackened rocks that he said was the result of a lucky grenade throw form the Japanese troop. Everyone could understand why the soldiers did not climb to confirm Murray's death. Murray had slipped into the cleft in the mountain and had survived the onslaught. By not returning fire he had assured his safe outcome.

Hearts were beating fast. Murray had a great uneasiness attacking his heart. He wondered if his defibrillator would go off. He had not been this nervous since the angel Colette treated him for his heart failure. Would he see the mystery or just confirm he had been hallucinating? The cave opening was very narrow. It looked like a simple crack but was large enough for an adult to squeeze during expiration. Tracie, who was slightly more curvaceous than Collette and had some difficulty wedging her hips through the opening. Gentle tugs from Colette and a push from the Houckster freed her from embarrassment.

The passage way was tortuous but gradually got larger and signs of habitation replaced the natural irregular cave walls. The light was dim from their lantern, but in front of them was a shiny object. The object was 7 feet tall and very irregular in shape. The top was flat like a table and trifurcated into three trunks. A bag like growth protruding from the lateral wall obscured one most anterior trunk. A snake appeared to come from under this growth and traveled in a serpentine manner to the base. The base was narrow and appeared to be floating in space. The object looked a little like a football with three sawed off trees growing from the top. The object was tarnished with glimmering shinny scales where age had not caused oxidation. Around the walls were the very silver objects that Murray had hallucinated. They appeared to be animals of four different types. Some of the animals were not recognizable. There were figurines of people. Every race, size and body shape was represented. The figurines were placed around the semicircle in an undulating manner. There were rooms visible from the center of the chamber. A system of lighting was arranged around the wall and to the foursome's amazement there was oil to feed the wicks. The room was enlightened with a single match.

The Houckster began to hyperventilate and fell to the floor. Tracie and Colette had immediate panic, never seeing the Houckster in such an undignified condition. They ran to his side and wondered if there had been a booby trap that injected him with silly solution. The Houckster could not speak and only could point to the strange football in the center of the room. When the Houckster slowed his respirations and returned from his near faint he said, "Can't you see? Are you blind? Have I failed in my teaching?

Murray responded, "Slow down young doctor. If you recognize something in this puzzle spit it out."

Collette was now staring at the central object. A sudden realization fell across her face. "A heart, it is a heart; it is a 'fricken' giant silver heart."

Closer examination of the heart revealed great detail. It looked like a 64 slice CT three-dimensional rendition of the heart. The superior vena cava, pulmonary artery and the aorta were the trunks. The most anterior trunk was the pulmonary artery and anterior to it laid the left atrial appendage the bag that extended from the lateral wall. The snake that was crawling out from under the appendage was the left anterior descending coronary artery. Diagonal blood vessels, the circumflex with obtuse marginals and the right coronary artery were clearly visible as you walked around the heart. The right ventricle and atrium along with the right atrial appendage were visible. Behind the heart four vessels were seen entering the left atrium. When the light was positioned just right there was a sense that you could see into the silver object and see the valves and subvalvular apparatus.

The Houckster was in awe. He had studied the heart for decades and here in the middle of the jungle was a perfect model that was centuries – or a millennia old. At the bottom of the heart were symbols bound together like a hieroglyphic. The meaning of these symbols would have to wait. Feeling across the metal surface of the heart he found thin cables that were placed on the lateral, inferior, and apex of the heart. The cables traveled to small golden box. PACEMAKER WIRES AND A PACEMAKER!! The Houckster thought he had contacted jungle fever and was hallucinating. If this was an ancient artifact, the civilization was far advanced.

The Houckster took a breath and then commented, "I know we could spend hours looking at just one object. We should begin by making a general tour and try to decide the general theme of this place. We must unravel the fantastic story this archive is trying to tell. Tracie, I hope your digital camera is charged. Murray, I know you brought a 35 mm. Let's begin by standing back surveying each room and taking as many photographs as we can to help record our findings."

Going back to the entrance and looking at the heart from the entrance. It became obvious that the heart was in the center of a curved galley. On the walls were curving shelves with figurines spaced at precise intervals. The shelves on the walls gave the appearance of moving in and out of the wall in a slow spiral. The shelves were paired and figurines were paired. The figurines were of 4 varieties. This backdrop behind the great silver heart was familiar. Double stranded DNA, a structure first elucidated by Watson and Crick came to mind.

DNA is the secret of life. DNA is the Turing Machine of the body. This protein is the road map to cellular function and ultimately of the human body, which is an assembly of cells. The DNA protein is the messenger between generations and transmits to off spring certain characteristics. Mendel in 1865 first noted that certain characteristics of a plant's parent were transmitted to its off spring. He used garden peas with different characteristics and cross-pollinated these plants to make tall or short plants. He worked out systems to predict the out come. Mendelian genetics is still taught today. These genes determine the type of plant that will develop from seed. The genes were strung together in chromosomes, which lived in the nucleus of the cell. The chromosome was composed of DNA. The discovery of the code of life was from the contributions of many: Rosalind Franklin, Erwin Schrödinger, and Oswald Avery. Breaking the code is still an ongoing effort. Most genes simply are a code for protein development. If there is a mutation of the gene, either the protein may not be produced, or the protein produced is abnormal and no longer performs its intended function. Mendelian genetics revolves around the simplest of changes single gene mutation. These single gene mutations result in a disease state. One in 200 live births has a single gene mutation. Some of these diseases are sickle cell anemia, hemochromatosis, Marfan's syndrome, and cystic fibrosis. Single gene disorders can be X-linked so that only a certain sex of an off spring is affected, autosomal recessive or autosomal dominant determine the probability of having the disorder.

Coronary artery disease is a polygenic disorder meaning there is more than one gene involved. There may be interplay between genes and the environment. Other examples include most of the dreaded diseases including breast cancer, Alzheimer's, diabetes and obesity. The genes that determine your individuality are polygenic and determine hair color height and allows for no duplication from separate births.

Chromosomal abnormalities occur when whole sections of the genetic code are deleted or duplicated. Down's syndrome is an example. These abnormalities are more easily identified since separating the chromosomes and viewing under a microscope will identify those changes. Mitochondria, or the powerhouse of the cell, has its own DNA. Mutations in this DNA will affect the function of mitochondria.

The Human Genome Project was begun in 1990 and completed in 2003. The US Department of Energy and The National Institutes of Health coordinated the study.

Partners in the project included many countries and significant contributions were made by private industry. The goal was to determine the sequence of the 3 billion base pairs and locate and identify the estimated 25,000 genes that determine life's characteristics. There are good genes that can resist disease and bad genes that may promote disease. An example, given by Dr. Robert Roberts during an interview on the ACCEL Tapes is the transmission of the AIDS virus. In Caucasians after an inoculation with the virus, AIDS is only transmitted 70% of the time. Some Caucasians have innate immunity. In the African-American population the transmission rate is nearly 99% with no immunity. This disease is ravaging the continent of Africa.

Dr. Roberts is using a shotgun approach to find polygenetic abnormal genes that cause coronary disease. In all of the 3 billion base pairs there is less than .1% variability that determines all of your characteristics. The monkey and humans are 98.5% identical. Dr. Roberts plans to compare the 3 million base pairs that are different between young individuals with and without coronary disease based on a 64-slice CT scan of the coronaries. He will have to map the entire human genome in 4,000 patients and compare the differences in the 3 million base pairs. Once he has found the differences he will then have to determine the function of that gene by placing it into some cell and see how the function of the cell changes. Considering it took thirteen years to map the first human genome, it appears his task is impossible. Technology has helped speed the process. What we will do with the answers that are on the horizon is still undetermined.

Colette broke into the lecture on genomics, proteomics, and cellular therapy. "Murray, you were here in 1944 or 45. How could this stuff have existed if it wasn't even invented?"

I had a concussion and was ill with fever. I was just happy to be alive. I did not know what my eyes were recording but I had a sense that this place was not primitive. The myth of UFO's was not even invented. This place seemed to be out of time but I had no background to make an assessment. The recurrent dreams of this place made me search for answers over the next 60 years. I would read about rumors of visitations from beyond our world, or ancient civilizations that existed and vanished without a trace. Archaeology raises more questions than gives up answers. The most outlandish of the rumors states there were four different advanced civilizations on the earth and that they were linked to each other. One civilization looked to the heavens for answers, one looked to the sea, one looked to the land, and one looked to the secret of life. Sitting here across from the great continent of Asia is repository of the civilization that wished to unlock the mystery of life.

The Mayan culture disappeared from the South Western Hemisphere may be the closest example of a culture that looked to the skies. Some theories imply they came from the sky. Their repository may have been destroyed and the only relics of this culture were four painted texts. Do these texts suggest a myth or a legitimate warning of the end of the world, as we know in 2012? Somewhere in the Atlantic the sunken city of Atlantis may hold the secrets of the sea. The civilization of the land is the least known. It is a good bet that it is buried under Antarctica or even hidden on the dark side of the moon. I suspect they were the terraformers who could balance carbon and water and energy to make a habitable world. References to each other may be in this room or in one of the attached rooms.

Silence filled the room, until the Houckster again took charge of the investigation. He did not think much about leadership, but a good mystery, a puzzle to be solved was enough motivation. An uneasy feeling from the "force" was also a motivating his effort. He could not explain the feeling he had in his stomach. Discovering this place was enough to give him a nervous stomach. This place had been undiscovered for centuries, but he had the feeling their time to explore was short. There were three additional exits from this room. To save time, he suggested they split and explore two of the rooms and meet in the last room. They were to look, take pictures but not touch anything.

It appeared they were in a museum, but if the ancient culture knew about pacemakers there could be many more surprises. He thought about the story Murray had just told. It seemed true relating many ancient mysteries to one source. The earth, sky, and water were essential components of their world. They followed physical laws and were subject to entropy the second law of thermodynamics. Everything in the sky, land, and sea became more chaotic in time. Mountains would erode. The sea's composition changes as the land would dissolve into the water. The air would fill with changes from great large volcanoes and from tiny plants that would process the gas from the air and even human intervention. There were cycles but in the natural process everything got more disordered. Life was different. Edwin Schrödinger, a physicist turned biologist, stated that the definition of life was negative entropy. Metabolism was a means of turning energy into an ordered existence. The earth was always a battle between positive and negative entropy. Aging was simply defined as an increase in entropy with more chaos. The Houckster looked at Murray and observed the miraculous changes. He had used energy to combat aging and ordered his life instead of allowing it to deteriorate.

The foursome split into two search parties. Murray and Colette moved toward the first passageway to the right and Houckster and Tracie went to the middle passage. They would meet in 15 minutes on the far left passage that was closest to the caves entrance. A single whistle blow would denote that one party had finished. Two whistles would denote come and see and three would be come there is a problem.

Murray and Colette moved through the passage and ignited the lamps. Murray immediately recognized the great wheel of the Mayan calendar and took pictures. The wheel was in the center of this room with a small inscription at the base. Murray recognized the symbols that correlated with the date 21 December 2012. Along with this date was a reference to the Heart seen in the main chamber. The reference was by means of a small heart model. This room was elliptical and presented the planets in near field and Milky Way in mid-field and other galaxies in the far field.

Tracie and the Houckster entered the middle chamber and in the middle of this room suspended in mid air was the molecule H_2O. Around the room were small figurines that depicted everything from tiny organisms to great whales. There were diagrams that represented the great currents of the oceans and what Tracie interpreted as the cycle of water. At the base of the H_2O molecule there was a gleaming city. They were just beginning to take pictures when the first earth-shaking tremor occurred with nearly simultaneous whistle blasts. The four nearly ran into each other as they exited their respective rooms blowing the whistle in a continuous tone. The Houckster dove to the base of the heart and snatched a parchment. The four ran to the exit in a hurried but orderly manner and the three pushed Tracie through the narrow opening. Murray exited the cleft just as the second tremor closed the opening and sent a tumultuous rock fall onto their heads. The ground was now bobbing like a cork in a bathtub.

The earthquake and aftershock had lasted less than 45 seconds. The realization that they were alive was still forthcoming. Cuts and bruises were evident but no injuries appeared life threatening or incapacitating. They had just realized how lucky they were when Murray cried out and slumped to the ground nearly flinging him off the crumbling ledge. He had five more shocks from his defibrillator until the beast in his chest rested. He gave an un-intended cry with each shock. He never lost consciousness. Balancing on the ledge and not falling the fifty feet was not an easy task.

The Houckster took his pulse for more than a minute and then looked down at Murray and inquired. "I did not think anything scared you."

"Scared?, Scared,? You bet I was scared!. I was the last one out of the cleft and could feel the mountains sliding together. I thought I was going to be caught half in and half out. I am even more frightened after the mule kicked me six times. What happened? "

The Houckster who was by his side taking his pulse replied, "Your heart rate is 140 and slowly decreasing. I suspect that your defibrillator thought you were in ventricular tachycardia when you were merely scared to death. Sinus tachycardia is normal when you think you are dying. It is also likely you did not take your medications this morning because you were up before breakfast, so you had no protection from your beta blocker. Murray you are going to be OK."

Murray looked up in a half grin and gurgled a reply, "Doc you are always on the same band wagon. Take your medications, take your medications, and take your medications. You have that cute Tracie trained like a parrot and that is the only thing she says. Even Collette has been poisoned by your incessant preaching."

The foursome looked around and took stock of their situation. The cleft was closed and they were on a 3-foot ledge that had lost several sections on either side. The first order of business was to get to lower ground away from a potential landside from potential aftershocks. After adjusting the repelling anchors they made it to their base camp within thirty minutes. They had broken camp and were walking away from the hillside when the mountain began a thundering collapse. The ledge was gone and debris fell to within 50 feet of their camp. They were all glad that they had moved in a hurry.

The next three days were fatiguing but uneventful. They gave away their camping gear for a ride to Manila in a Jeepney. Two flat tires later they were in a five star hotel soaking in their respective baths. They did not discuss their adventure, but rather just enjoyed each other's company. The plane ride home was scheduled for the morning.

Three months later, the grill for Mongolian Barb-B-Q was heating. Murray, Tracie, and Colette were sitting around the pool while their host the Houckster and his wife prepared the vegetables for the grill. The photographs that they had taken were displayed. They had all shared their copies and were trying to make sense of the pictures. The pictures did not represent what they thought they had seen. Without a panorama and feeling of depth, many of the objects and symbols were simply mysterious. Unraveling the meaning was difficult. Murray was successful in the Sky Room finding symbols that were similar to the Mayan calendar and confirmed the date of the ending cycle. Strong reference was made to the life room. Murray had concluded that the end of a cycle referred to a biologic event and not due to a catastrophe from the sky.

The Houckster pulled out the parchment that he rescued from under the great heart. Surprise rose in the eyelids of his partners. They were unaware of this document, the only relic from their adventure. Pointing a single diagram he simply said telomere.

December 21, 2012, co-incidence with the end of a Mayan cycle, a single cell was injected into an aged mouse and placed upon a treadmill. There was no galactic tragedy, major war, or environmental upheaval. The end of the Mayan cycle was not even noticed. The revolution to come began with a small insignificant creature. The cell had been harvested from the mouse two weeks earlier. Over 300 million cells had to be searched for this pleuripotent cell. The cell was then injected with virus that had simple instructions. The instructions were to increase the length of the telomere by 10 times its dwindling size. The telomere was the counter for cell division. By lengthening the telomere the cell would continue to divide. It was the Turing machine that determined the natural lifetime. The mouse stem cells were nearing the end of their ability to divide. When the stem cells were no longer available to replace senescent cells natural life would slowly grind to a stop.

Gene therapy was initially tried in genetic disorders where a protein was missing. It was thought that every cell in the body had to be infected with a safe virus. The virus would be the Trojan Horse that would repair the DNA that coded for the protein. The initial experiments were a disaster with the recipients dying from the viral infection. It wasn't until cellular therapy was applied to genomics and proteomics that a solution could be found. A pleuripotent cell had to be found. This cell then had gene transfer to modify this single cell's DNA code. The cell was then observed as it divided and differentiated in the cell culture. The offspring were observed for any deficiencies and to see if the missing protein was now present. After multiple safety checks, and multiple divisions of this single cell, the harvested cells would be placed into the patient through a simple intravenous injection. In four months these new cells would be rejuvenating organs and going about repair. Eventually the genetic deficiency would no longer be present affecting cure. Pleuripotent cells have the ability to divide. Even these cells have a finite number of divisions that determines their lifetime. The telomere was the measuring stick. By lengthening the telomere life could be extended.

It only took a single pleuripotent cell to be rejuvenated to repair the aged mouse. The other ingredient required was exercise to mobilize these stem cells from their resting place in the bone marrow and scattered along the endothelium through out the blood vessels. Exercise was the key to mobilizing this engineered cell. Over the next three months, the mouse grew stronger, gained weight and lived more than three times its natural lifetime until necropsy was performed to determine if there were any abnormal cells lurking in its organs.

The puzzle of the cleft was the knowledge of life extension. The end of society and upheaval of many cultures was predicted. The astronomical date of this successful experiment just happened to coincide with December 21, 2012. The biggest catastrophe was not a deadly comet, an aberration of the carbon cycle. The biggest catastrophe now was a conflict between young and old and who would receive prolongation of life.

Heart failure and vascular diseases are deadly. However, there is hope. Treatment can restore individuals to good health and help to maintain health. The best treatment is prevention by good living habits and faith. I hope *Take Heart* will guide you in the proper path and help you to guide others in this path.

Retirement does not mean it is time to sit back and let society care for you. It simply means you have changed your occupation and your goals are now different. Your new goals include looking out for your fellow man. Care for them as they may have to care for you. Baby boomers become a resource and not a drain.

The following are key points I would like to stress from *Take Heart*.

Exercise is a lifetime prescription.

Houckstersize your restaurant meals.

Life extension is science fiction but not unreasonable. Live like it is a reality.

Nursing home and mortality issues should be part of family discussions.

Medicine can be sad, rewarding, and demanding. It is a worthy and fulfilling occupation.

Question your teachers and your leaders.

Take responsibility for maintaining personal medical records.

Use democracy to shape our delivery system of health care and change the future of medicine. The entire Earth will benefit.

Question your teachers.

I may have to write a book to give this simple thought notoriety and gain public support. I will develop a wellness web page that will determine risk and suggest methods to lower risk. The web page will be called http://houckstertakeheart.com/. It will evolve into an interactive medical record providing feedback to the individual and to the system of medical care placing the responsibility of health on the individual!

Glossary

12 lead electrocardiogram	Dr. Wilson standardized the electrocardiogram with 6 precordial leads and three limb leads (Eintoven's triangle) that could be combined in 6 ways for a total of 12 leads. Each lead has a different electrical view of the heart. The skin should be sand-papered for good contact.
150 Joules	The upper limit of electrical energy delivered when a patient is shocked by a defibrillator .
65u (65 micros)	Measure of length about the thickness of a human hair.
750 cc	Cubic centimeters is a volume measure and 750 cc's is 0.79 quarts.
ACCEL	Educational cardiology tapes sponsored by the American College of Cardiology. An excellent means of staying current by listening to tapes while you drive to work.
ACE-Inhibitors	A hypertension medication that inhibits Angiotensin conversion to a potent vaso-constrictor.
Acute coronary syndrome	An unstable condition caused by the formation of a clot inside the coronary artery. Chest pain and shortness of breath are the primary symptoms and if the clot completely forms, the result is myocardial infarction and possibly death.
acute myocardial infarction	A serious life threatening condition that occurs when a clot forms in a coronary artery resulting in heart cell death. If a vessel is opened in less than an hour, little permanent damage results. The longer the blood vessel stays closed, the worse the outcome.
adenoma	A small tumor producing a hormone.
ADP	Adenosine Diphosphate is a nucleotide that is packaged in platelets.
AFCAPS	Air Force/Texas Coronary Atherosclerosis Prevention Study is the first primary prevention trial using the statin Mevacor showing cardiac events and strokes were reduced with this medication.
afterload	The size of the arterioles, the resistance to flow from the heart.
Aggrastat	An IIb IIIa inhibitor of platelets that lost market value because it did not use a large enough loading dose.
AICD	Automatic Implantable Cardioverter Defibrillator
alcohol ablation	Injection of alcohol to kill tissues.
Aldosterone	Adrenal hormone that is involved in salt and water balance within the body and is responsible for inflammation and fibrosis.
Algorithms	A predefined procedure based on input variables to help make decisions.
alpha blockade	An inhibitor of the sympathetic constrictor neurohormones .
analog	Similar to another protein or molecule.

angina	Chest discomfort similar to a burp that is stuck or swallowing ice cream to fast. It is a heart pain that is related to exertion.
angiogram	The pictures of blood vessels and cardiac chambers with the use of a dye that absorbs radiation. See Angiography.
angiography	The method of taking pictures of blood vessels and cardiac chambers with the use of a dye that absorbs radiation. The pictures can then be seen on X-ray film or by digital detectors. Similar terms are cardiac catheterization, cath, angiogram. All of these terms mean a needle puncture of the artery to gain access to the vessels that will be imaged.
angioplasty	The process of intervening on a blood vessel to change its shape. Initially performed by the Dottored method of dilatation and later use of balloons and stents. This technique will smash a blockage into the vessel wall to make the lumen of the artery bigger.
angioplasty sheaths	The IV that is placed in the groin artery so the catheters and angioplasty balloons can be passed into the artery to the heart.
ankle brachial index	The ratio of systolic blood pressure in the ankle divided by the systolic blood pressure of the arm - greater than 1 is normal, less than 1 suggests peripheral vascular disease and a greater chance for heart attack and stroke.
anterior distribution	The blood supply to the front of the heart supplied by the left anterior descending and diagonals.
Anti-arrhythmic medication	Drugs that will reduce irregular or dangerous heart rhythms.
anti-coagulation	The prevention of coagulation - the clotting of blood. Drugs known as anti-thrombins such as Heparin or Lovenox are given to prevent clots from increasing in size in blood vessels and the heart. Other anti-coagulants are in development. The major risk of anticoagulation is bleeding.
Anti-coagulation therapy	Drugs that inhibit the clotting system.
antioxidant	A drug or vitamin that shields cells from damage due to free radicals. They are able to nullify the extra unpaired highly reactive electrons on free radicals.
Anti-thrombotic	Prevents clotting.
aortagram	The process of taking pictures of the aorta the main vessel leaving the heart to the body. This done with a power injector to deliver 60 cc's of contrast dye into the aorta in 2.5 seconds.
aortic insufficiency	Leak of the aortic valve the main valve out of the heart to the body.

Aortic stenosis	The aortic valve becomes stiff fused and can not open resulting in a pressure gradient across the main valve from the heart. This is like putting your finger on a hose and making it squirt with more force.
apoptosis	Cell death.
arrhythmia	Not in rhythm. Some arrhythmias are benign, some cause symptoms such as weakness and palpitations, while others are dangerous.
arterial vasodilator	increase the size of arteries so more flow can pass through the vessel.
arterioles	Small arteries.
astigmatism	Changes in curvature of the lens or shape of the eye resulting in blurred vision.
Asystole	No heart beat, flat line, death ensues if a heart beat does not return.
atherosclerosis	The process of fat, cholesterol, and inflammatory cells building up inside of arteries with blockage of blood flow.
atherosclerotic plaque	A small bubble inside the artery that protrudes into the flowing blood. The shell around the bubble is a fibrous cap. See plaque.
Atrial	Refers to the upper collecting chamber of the heart. There is a left and right atrium.
Atrial septal defect	A hole between the left and right atrium. This is the most common congenital heart defect in adults and allows flow to shunt from the left atrium to the right. This can be benign, result in pulmonary hypertension, or occasionally in a paradoxical embolus when a blood clot from the vein passes to the brain.
Atrial-ventricular node	Specialized electrical conducting cells that slow conduction of electrical impulses that are traveling from the Atrial fibers to the ventricle. This allows the atria to fill the ventricle priming the pump before the ventricle contracts.
AV nodal ablation	Destruction of the atial-ventricular node causing an electrical blockage between the upper and lower pumping chambers.
AV valve	The valves between the atria and ventricle mitral on the left and tricuspid on the right.
back up defibrillation	A pacemaker or device that monitors the heart for lethal rhythms and delivers a shock to prevent death.
basal	Resting while awake but not sleeping.
basal metabolic rate	The energy used when resting not sleeping and not working. If you do nothing it takes very few calories to keep you alive.
bathysphere	A special thick walled submarine that is used to explore the ocean's depth.

beta-blocker	A drug that blocks the beta sympathetic receptors. They prevent the adrenal hormones from increasing stress heart rate and blood pressure.
bicuspid aortic valve	Aortic valve with two leaflets instead of three. The most common congenital heart disease in 5% of the population.
BiDil	A combination pill of Hydralazine and Isordil Dinitrate that is over priced.
bigeminity	Every other heart beat is early premature. Sounds like: dah dah dah dah dah dah.
biphasic energy	The energy delivered by modern defibrillators delivers positive and negative voltage in sequence and is more effective than pure positive or pure negative volts.
bipolar disorder	A disorder of brain chemistry that makes you very happy (mania) or very sad (depression) causing atypical behavior patterns.
biventricular pacing	A pacemaker that activates both the left and right ventricle and helps improve cardiac dys-synchrony.
BNP	Brain Naturetic Peptide is a substance produced by heart cells when they are under stress.
bolus	A rapid infusion of a fixed amount of drug or fluid.
brachial vein	A vein found between the arm and forearm anterior to the elbow where blood is commonly drawn.
bradycardia	Slow heart rate below 60 beats per minute.
bradykinin	A peptide that causes vasodilatation, cough, and inhibits cell death.
bronchiectasis	Abnormal destruction and widening of airways that results from and causes infections of the airways.
bruit	A *Shssssss* sound made by an artery that has a narrowing due to abnormal blockage.
BUN	Blood Urea Nitrogen a marker of kidney function
C reactive protein	A protein made in the liver when there is inflammation in the body. See HS CRP and CRP.
CABG	Coronary Artery Bypass Grafting is a surgical procedure that splits the sternum and provides a bypass around a blockage in a coronary artery.
canon A waves	Periodic large waves in the jugular vein when there is complete heart block or ventricular tachycardia causing AV dissociation.
canula	A tube such as oxygen tubing that hangs off of the ears and supplies oxygen to the nose.
capillary electrometer	A delicate tool to measure small changes in voltage and was used to record the first electrocardiogram.
CAPRICORN Trial	A placebo controlled beta blocker trial with Coreg in the treatment of myocardial infarction performed with potential early harm and late benefit and questionable ethics.

cardiac dys-synchrony	If your heart was like a V-8 Engine cardiac dys-synchrony would be like having two spark plugs pulled off with the heart running rough and inefficient.
cardiac enzymes	Blood test that looks for evidence for myocardial cell death. When cells suffer an injury these enzymes leak into the blood and can be used to diagnose a myocardial infarction.
cardiac resynchronization	(See cardiac dys-synchrony) Resynchronization is a process where additional spark plugs are placed on the V-8 engine heart to make it run better. This can be accomplished with special pacemaker wires.
cardiogenic shock	Low blood pressure with congestion due to severe heart dysfunction.
cardiology fellow	A young doctor who is certified in internal medicine and is training to become a cardiologist. The staff cardiologists break them down and rebuild them into superb brilliant and hard working doctors.
cardiomyopathy	Sick weak heart muscle or infiltrated heart muscle resulting in poor pumping.
carotid artery	One of the first major arteries off of the aorta and supplies blood to the right and left brain.
carotids	See carotid artery.
Carvedilol	Generic for Coreg - A beta-blocker with alpha blockade and antioxidant properties that is an excellent drug for heart failure.
CASS trial	Coronary Artery Surgery Study clinical trial.
catecholamine	Adrenal hormones that increase contractility of the heart, raise heart rate fight of flight hormones.
catheterization	(See angiogram.) It is also the method of measuring the pressures and flows within the chambers of the heart.
CCU	Coronary Care Unit was developed in the late 1950's to monitor patients with myocardial infarction.
central lines	An IV that is placed into a major vein usually in the neck, groin, or under the clavicle.
CHF	Congestive Heart Failure has many faces and means congestion of the lungs due to poor cardiac function.
chordae	Small collagen fibers that attach the mitral valve leaflets to the papillary muscle.
circumflex	One of the three coronary arteries. It branches off the left main and travels in the AV groove to supply the lateral wall of the heart.
cirrhosis	Fibrosis of the liver with poor liver function alcohol is one etiology for this condition.
claudication	The cramping pain in the calf or buttocks that comes on with walking. The cramp is due to inadequate blood supply.

Co Q-10	An anti-oxidant vitamin that is the final protective pathway against free radicals.
coapt	To close in a uniform manner forming a tight seal.
code	The process of resuscitation of a patient from cardiac arrest Code Blue.
collagen	Strong tuff bundles of protein fibers that give structure form and allows attachments to other structures.
complex ventricular ectopy	Frequent irregular beats that originate from the ventricle.
congenital heart disease	The abnormal structural development of the heart resulting in shunts stenosis and malformations that may need to be corrected or palliated by surgery. 0.8% of births may have some structural abnormality.
congestive heart failure	Congestive heart failure has many faces and means congestion of the lungs due to poor cardiac function.
Conn's Syndrome	An Aldosterone producing tumor of the adrenal gland.
contrast media	This is a solution that is injected so vessels and chambers of the heart can be visualized. Contrast media for X-rays has a high Iodine content and can be damaging to the kidney. Although rare, an allergic reaction can sometimes occur as a side effect.
Cook needle	A bevel tipped hollow needle that is about 2.5 inches long and is used to puncture the femoral artery. It is almost too short for today's heavier patients.
coronary angiogram	Cardiac Catheterization, Cath, Angiogram is the method of taking pictures of coronary arteries with the use of a dye that absorbs radiation. The pictures can then be seen on X-ray film or by digital detectors. All of these terms means a needle puncture of the artery to gain access to the vessels that will be imaged.
coronary angiography	The technique of taking pictures of coronary arteries with the use of a dye that absorbs radiation. The pictures can then be seen on X-ray film or by digital detectors. Similar terms are cardiac catheterization, cath, angiogram. All of these terms means a needle puncture of the artery to gain access to the vessels that will be imaged.
coronary atherosclerosis	The build up of blockage in the coronary artery. It occurs at branch points and areas of mechanical stress. It is composed of a fibrous cap that contains cholesterol and inflammatory cells. Men have more proximal blockage and women have more distal blockage. Blockage is bad.
Coronary Artery Bypass Surgery	Coronary Artery Bypass Grafting is a surgical procedure that splits the sternum and provides a bypass around a blockage in a coronary artery. It can be performed on and off the heart lung machine. The second operation is not as easy.

Coumadin	Trade name for rat poison, it is the oldest oral anticoagulant.
CPAP Machine	A machine that is attached to a mask and blows air into the patient (positive pressure). It is commonly used at home for sleep apnea and is used for patients with breathing trouble.
creatinine	A breakdown product of muscle that is cleared from the body by the kidney and can be used as measure of kidney function.
cytokine	Small signaling proteins that allow inter cell communication and are responsible for managing inflammation as well as other destructive and reparative activates.
DAVID trial	The dual chamber and VVI implantable defibrillator trial. This matched the best versus simple and to the surprise of all less technology won. This demonstrated that pacemakers could cause cardiac dys-synchrony.
decubitus	To lay down, also refers to bed sore or pressure sores.
defibrillator	A special form of a pacemaker that monitors for bad rhythm and delivers a shock to cure the arrhythmia. Defibrillators can also be eternal devices that are placed on the chest by a friendly bystander. These are known as AED Automatic External Defibrillators.
Dengue Fever	A hemorrhagic fever low platelets bone break fever transmitted by mosquitoes.
depolarizing	A decrease in the voltage across a cell membrane. This occurs during systole and controls the contraction of the heart or skeletal muscle.
desquamating	To flake off.
diabetes mellitus type I	Usually juvenile in onset with an inability to regulate glucose due to a lack of insulin and these individual will die without insulin replacement.
diabetes mellitus type II	Adult onset glucose insulin disorder that allows sugar level to climb. Patients will not die with out insulin but will have frequent urination and die of vascular disease.
diaphoresis	A cold, perfuse, drenching sweat when one is sick.
diastole	The filling phase of the heart. The time the heart is recovering after a contraction. The time from Aortic valve closure to aortic valve opening ignoring the isovolumic phases.
Dig trial	A trial to determine if the oldest medicine for heart failure is safe and effective. It did not change mortality but did result in less hospitalization.
Digoxin	The oldest medication for heart failure and is still useful despite recent de-emphasis by current guidelines.
Dilaudid	A frequently abused pain killer.

Disclosure	To tell of your financial interests.
dissection	Splitting of the inside of a vessel so the inner lining flaps in the lumen.
distal circulation	The end or terminal portion of the blood vessel.
distal tubules	Part of the nephron that collects urine and regulates salt and bicarbonate reabsorption.
distal vessels	See distal circulation.
diuretic	A medication that stimulates urine production and usually works by excretion of sodium.
diuretic	A water pill.
dobutamine echocardiogram	A resting stress test for individual who can not walk. The heart rate is artificially increased using adrenal hormones and the wall motion of the heart is observed with an echocardiogram. The patient lies on the table and does no work.
dual chamber pacing	A pacemaker that has a lead in the upper chamber right atrium and in the lower chamber the right ventricle. It takes two leads to simulate physiologic contraction when the AV node no longer works properly.
Dura	Lining of the brain.
dyskinetic	To move in the opposite direction an aneurysm is a dyskinetic segment.
dyspepsia	Heart burn.
dyspnea	Shortness of breath.
Echocardiogram	An imaging machine that takes pictures of the heart with the use of ultrasound. The ultrasound can penetrate deep into tissue and the reflected echos can be recorded and turned into pictures. It is similar to the machine that takes pictures of babies in the womb.
echocardiographic	See echocardiogram.
ectopy	Additional, or early heart beats.
edema	Swelling of the distal extremities due to excess salt and water. Swelling can occur in the lungs and cause shortness of breath or in the abdomen and cause ascites.
EECP	See Enhanced Extracorpeal Counter Pulsation.
Ehrlos Danlos Syndrome	"Indian rubber man, contortionist." An inherited disease of collagen that makes one flexible because of abnormal collagen.
ejection fraction	This is the measure of the squeezing function of the heart. It is the amount ejected (end diastolic volume - the end systolic volume) divided by the end diastolic volume. A normal ejection fraction is 60%, after an inferior MI it will fall to 40-45% and after an anterior MI 35%. The lower the ejection fraction the worse the prognosis. For the very body ejection fractions the bigger the heart the worse the prognosis.

electrical remodeling	Changing the structure of the heart through alteration of the electrical conduction or stimulation.
electrolyte	The class of cations and anions that are in the blood and cells sodium, chloride, potassium and bicarbonate.
electrophysiologic testing	Pacing the heart in a study to produce arrhythmias or death.
embolus	A clot, vegetation, or piece of natural or man made material that travels in a blood vessel and occludes the distal circulation.
end hole catheter	A catheter or long thin tube that has only an end hole and no side holes.
Endocannabinoid receptor blocker	A medication that blocks receptors in the body and brain that marijuana tends to stimulate.
endocrine axis	Glands that produce hormones, their targets, and feedback loops to control hormones.
endothelium	The specialized single cell layer that lines all of the blood vessels. It is metabolically active and controls repair of tissues, clotting, and inflammation, and controls nutrients and vasodilatation of vessels. It is the only substance that can contact blood without clotting. It is the largest organ in the body (it really is not an organ but I wish to give it this status)
Enhanced Extracorpeal Counter pulsation	A machine composed of a blood pressure cuff in the ankles, legs, and hips that inflate in sequence when the heart is resting (diastole) and pumps blood in the opposite direction toward the heart. It is used in the treatment of angina and heart failure.
EPHESUS trial	A trial of a Spironolactone like agent that does not make breasts tender, Eplerenone was studied in post myocardial infarction with heart failure.
epicardial coronary artery disease	Blockage of coronary arteries in the proximal segment on the out side of the heart.
epicardial vessels	The coronary arteries that are on the outside of the heart the epicardium (around the heart).
Erythropoietin	A hormone that increases red blood cell production.
evidence based medicine	Practice of medicine based on outcomes from properly performed clinical trials (experiments).
fatty acid pathway	The way fats are broken down and metabolized.
femoral artery	The major artery to the leg. It can be palpated in the groin half way from the hip bone to the centerline of the body.
fibrosis	The process of forming scar tissue.
Foley catheter	A tube that is designed to pass through the urinary system to the bladder. It can be difficult to find in some women and is difficult to pass in some men who have a large prostate. It is used to measure urine production closely and to keep the bed from getting wet.

Framingham study	An epidemiologic study was performed in Framingham, Massachusetts. Two thirds of the population agreed to the study and their survival was better. The purpose of the study was to find risk factors for vascular disease.
free radicals	Are molecules that are very energetic because they have unpaired electrons. They can be formed in the presence of oxygen. They tend to attach themselves and can damage blood vessels. The are blamed for atherosclerosis.
gallop rhythm	It is the summation of both the third and fourth sound in a tachycardic patient with heart failure.
Gastroesophageal Reflux Disease	Everyone has this condition after an overindulgent meal like Thanksgiving. 40% of people have a lazy esophageal sphincter (muscle band in the distal feeding tube) that allows acid and food to travel backwards into the esophagus. The acid will irritate the esophagus and cause a pain that is very similar to heart pain. All beds should be elevated 4 inches at the head end since most of the reflux occurs at night.
GCSF	Granulocyte Colony Stimulating Factor a protein that increases white cell production.
Genetech	A biotech company whose early reps make 10 times the amount of money that I make from stock options alone, because the company did a good job.
genomics	The study of all genes of an organism.
GERD	See Gastroesophageal Reflux Disease.
gynecomastia	Painful enlargement of the breasts.
HDL	Good Cholesterol greater than 45 in women and greater than 40 in men.
heart failure class	A classification of heart failure from I with no limitation to IV significant symptoms at rest.
heart lung machine	A machine that can replace the function of the heart and lungs for a brief period of time. It is composed of a pump usually a rotary pump and an oxygenator that will re-oxygenate the blood and remove carbon dioxide.
hemoglobin A1c	Glycosylated sugar on hemoglobin that implies diabetic control.
hemorrhagic stroke	A stroke due to bleeding into the brain usually resulting in death.
heparin	An intravenous drug that is used as an anticoagulant to prevent blood clots from forming. It was originally found in the liver and is purified from beef lung and intestines.
hepatic coma	When the liver fails to clear metabolic toxins ammonia levels rise and cause a coma.
High sensitivity CRP	A protein made in the liver when there is inflammation in the body demonstrating great efficacy in identifying individuals with increase risk of heart attack and stroke.

HMG CoA reductase inhibitors	See Statins.
holosystolic murmur	"Shssssss" sound that begins at s1 and continues through s2 throughout systole.
holter monitor	A device that records the electrical activity of the heart to detect arrhythmia.
homeostatic	The state that living organisms maintain to keep their internal environment stable.
Homocysteine	A naturally occurring amino acid in the blood. The vitamin B's and folic acid regulates production. It has been implicated in a cause of atherosclerosis. The inherited disease of Homocysteinuria has marked elevation of this amino acid and these individuals have severe atherosclerosis and thrombosis of their vessels.
homozygous	Having identical alleles. Alleles come from both parents and in some diseases if both are the same a disease process can be worse than different. The genes for a trait from the parents are identical.
Houckster	A handsome, George Clooney in appearance, intelligent, thinks out of the box, who is half Doctor, half Hamster. A Cardiologist who is hopefully a mentor to young doctors.
HS CRP	A protein made in the liver when there is inflammation in the body. HS CRP is a more sensitive test than CRP and has demonstrated great efficacy in identifying individuals with increase risk of heart attack and stroke HSCRP.
Hydralazine	A drug used in vasodilatation and as an antioxidant.
Hydrochlorothiazide	A water pill. See diuretic.
hyperdynamic heart	A condition when the heart has an ejection fraction greater than 60%. The ventricle is squeezing very hard and the heart is small. If the heart gets smaller with a prolonged stand a reflex will slow the heart to get better filling and the patient will pass out.
hyperkalemia	Elevated potassium levels.
hyperlipidemia	A condition where the circulating lipids are elevated. Lipids consist of the fat and cholesterol elements along with their transport proteins. Cholesterol total, HDL Cholesterol, Triglycerides are commonly measured lipids
hypertensive	Blood pressure elevated to high levels.
hypertensive heart disease	Thick stiff heart muscle from long standing hypertension that will eventually lead to heart failure.
hyperthyroidism	Elevated thyroid levels associated with tachycardia, weight loss, and Atrial fibrillation.
hypertrophic cardiomyopathy	Abnormal heart muscle that is thicker than normal.

Hypertrophic Cardiomyopathy with Obstruction	Abnormal heart muscle that is thicker than normal and also results in obstruction of flow out the aorta and induces mitral valve leak.
hypertrophied	Enlargement of heart muscle by increasing fiber size and increasing matrix. The thicker the heart muscle the stiffer it becomes. A weight lifter has hypertrophied skeletal muscle. A hypertensive patient has hypertrophied heart muscle.
hypertrophy	Enlargement of heart muscle by increasing fiber size and increasing matrix. The thicker the heart muscle the stiffer it becomes. A weight lifter has hypertrophied skeletal muscle. A hypertensive patient has hypertrophied heart muscle.
hypertrophy	To get larger. An increase in the number and size of cells.
hyperventilation	Breathing in excess of what is required. Carbon Dioxide levels will fall and the body will become alkalotic with tingling feeling of doom and eventual pass out spell. To control hyperventilation you need to rebreathe the carbon dioxide by breathing in and out of a paper bag.
hypokinesis	Normal function is now weak and is demonstrated as decreased movement of muscle.
hypotensive	Low blood pressure.
hypothalamus	Primitive part of the brain that links the nervous and endocrine system and controls the autonomic nervous system that regulates our sub systems.
hypothyroidism	Low thyroid with weight gain, fatigue, constipation and slow heart rate.
hypoxemia	Low oxygen levels in the blood.
hypoxic	Lack of oxygen, low levels of oxygen in the blood or tissue.
IIb IIIa inhibitors	Small molecules that compete with the IIb IIIa receptors on platelets and causes platelet dysfunction so clots can not form.
immune globulin	Antibodies collected from serum that can be given to another person.
inducible	In the electrophysiologist world they can make an arrhythmia occur
inferior distribution	The blood supply to the bottom of the heart supplied usually by the right coronary artery and one third of the time the circumflex.
inferior leads	Leads II, III, AVF on the electrocardiogram. These three of the 12 lead ECG look at the electrical activity of the bottom (also considered diaphragmatic) of the heart.
inflammatory markers	Biomarkers that can be detected in the blood that are elevated in inflammatory conditions. Examples include HS CRP and other cytokines.

inguinal	The groin.
Integrilin	An intravenous drug that is used to keep platelets from clumping, sticking together. This drug will help prevent heart attacks and has favorable affects on the repair of blood vessels.
Integrilin drip	An intravenous drug that is used to keep platelets from clumping. Drip refers to an old technique of adjusting the rate of infusion. The nurses would count the drops over a period of time and readjust to get the correct infusion rate. This is now done by machine.
interventional cardiologist	A cardiologist who places stents into coronary arteries, carotid arteries and leg arteries, closes holes in the heart by working through arteries and veins in the leg.
intra aortic balloon pump	A large balloon that inflates in the descending aorta during the resting cycle of the heart (diastole) and deflates during the pumping cycle (systole). The weak heart sees 40cc's of empty space and takes away work from the heart. Blood flow is augmented to the coronaries and to the brain. It is used for angina, heart failure, mitral regurgitation, and can not be used if there is aortic insufficiency.
intravascular ultrasound	A tiny crystal on the end of a wire that can produce and receive ultrasound. Using this technology, the vessel wall can be examined from inside and get a picture that is similar to a pathologic slide. It is used with angiography to help decide correct treatments.
intravenous lines	A catheter inserted into a vein to give medications or hydration.
introducer	A large IV that is inserted into the artery. Different types of catheters can be inserted and removed without having to stick the patient again.
ischemia	Lack of blood supply to a body part. Occasionally a excess of demand due to excess work over supply will also cause ischemia.
islet cells	Cell of the pancreas that make insulin.
IVUS	See intravascular ultrasound.
Jarvik pump	An artificial heart.
JL4 Judkins catheter	A catheter with two curves developed by Dr. Judkins (on cadavers) that will fit into the left main coronary artery. JL4 will fit most normal sized patients. If you are greater than 6 feet a JL5 may be needed, if you are short a JL3.5. The number tells you the size and L refers to a left coronary catheter.
jugular	The major vein from the head traveling to the subclavian. It is found in the neck below the triangle formed by sterno and cleido muscle bellies. There is an external and internal jugular.

Juvenile type I Diabetes	See Diabetes Mellitus Type I.
Kawasaki's disease	A desquamating skin rash associated with high fever and aneurysms of the coronary arteries.
Laennec	Dr Laennec died of consumption, (TB), described alcoholic liver cirrhosis, and invented the stethoscope.
LBBB	Left Bundle Branch Block. The AV node passes electrical impulses to the right and left bundle. A block of the left bundle implies no passage of electricity and results in cardiac dys-synchrony.
LDL Cholesterol	This is the bad cholesterol. It is calculated by subtracting the HDL (Good Cholesterol) and one-fifth of the triglycerides from the total cholesterol. A normal LDL is less than 70. The values in the blood depend on diet and LDL receptors. This bad cholesterol causes atherosclerosis.
LDL Receptors	The protein in the liver cells that attracts cholesterol attached to the protein LDL.
left anterior descending	The first coronary artery branch off of the left main and runs down the front of the heart. Obstruction in the proximal portion is nicknamed the widow maker because it has produced so many widows. In a jeopardy score it accounts for 6 out of 12 points.
Left Bundle Branch Block	LBBB refers to the AV node passing electrical impulses to the right and left bundle. A block of the left bundle implies no passage of electricity and results in cardiac dys-synchrony.
Leptin hormone	A protein involved in fat storage and dietary hunger and may be an explanation for some obesity.
level 4 four charge	A government term for the charge of a patient that must include appropriate documentation.
limbic system	The part of the brain and its connections that is involved with emotions and memory.
lipid solubility	Fat soluble and will cross into the brain since it is surrounded by fatty membranes.
lipids	Lipids consist of the fat and cholesterol elements along with their transport proteins. Commonly ordered values are Cholesterol, Triglycerides, HDL cholesterol. Lipid particles can also be identified by size electophorectic properties and by looking at milky serum.
lipoprotein	The proteins that attach themselves to the lipids and function to transport them to their storage space of help in breaking down the particles.
lumen	The hole in the vessel where the blood flows. The hole in a catheter. A cave could be considered a lumen of the mountain.

luminal irregularities	Evidence of mild blockage in vessels that do not impede flow but could be a site of a future heart attack.
lupus erythematosus	An auto-immune disease that attacks the body for no particular reason and results in inflammation of joints, skin, and inflammation of all organs including the heart.
LV function	Left Ventricular (left heart) function describes how well the heart is performing. Ejection fraction is one measure of heart function.
Maalox	Chalky, milky antacid.
macrophages	These cells have roles in the immune system and cleaning up debris in our bodies. They can become choked with fat and be a cause of atherosclerotic changes in our blood vessels.
magnetic resonance imaging	An expensive imaging machine that uses powerful magnets to align spinning electrons. It is similar to a CT Scan but does not use radiation.
mcg/kg/min	Micrograms per kilogram (body weight) per minute is a rate of drug infusion that is calculated according to a patients weight. Extremely overweight patients makes this calculation difficult because the distribution of the drug in a fat individual can be different in a normal sized individual.
medicated stents	Stents that have a drug coating to delay healing and excess cellular response inside blood vessels to prevent restenosis.
MEF2 gene	A defective gene on chromosome 15 that is involved with repair of arteries This explains why some heart attacks run in families.
meta-analysis	A study performed by lazy researchers who don't perform their own study but examine multiple similar studies and draw conclusions by methodically combining data from multiple studies.
metabolic syndrome	A collection of health risks that centers around obesity, waist size, diabetes, blood pressure , lipids and marks an individual who is likely to have a heart attack or stroke.
Metoprolol	A beta blocker medication . Ssee beta-blocker
Mevacor	One of the early Statins. See AFCAPS and Statin.
MI	Myocardial Infarction, better known as a heart attack, is the death of heart muscle due to lack of blood supply.
micro infarcts	Tiny infarcts due to swelling or obstruction of the microcirculation.
microcirculation	The circulation that is distal to large vessels and can not be seen with angiography.
mitochondria	Small organelles in cells that are the powerhouse making energy available for cell function.
mitral regurgitation	Leak of the mitral valve. The blood flow in the opposite direction into the left atrium.

mitral stenosis	The valve between the left atrium and left ventricle is thickened and does not open completely resulting in a pressure build up in the left atrium and lungs. The cause of this is due to rheumatic heart disease caused by group A Strep.
mitral valve	An AV valve that separates the left ventricle from the left atrium.
mitral valve leaflet	The valve between the left atrium and left ventricle has two leaflets. The leaflets are like flaps that close when the pressure in the left ventricle is greater than the atrium. The leaflets are attached to papillary muscles.
mmHg	A measure of pressure equivalent to the height of a column of liquid mercury.
modifiable risk factors	Risk factors that can be altered by change in lifestyle.
morbidity	Disease or disability due to a disease or accident.
mortality	Death.
mucomyst	Acetylcysteine is a sulfur containing drug for detoxification of Tylenol overdose and may protect the kidney from contrast mediated renal failure another name for this drug is mucomyst
mule kick	This is a self-explanatory term.
myocardial infarction	The death of heart muscle due to a lack of blood supply usually from an occlusion of a coronary artery.
myocardium	Heart muscle.
N of 1	An experiment with only one subject and is considered an observation and not good science.
Natrecor	Nessiritide, BNP, the perfect medicine to treat heart failure. It is a substance produced by heart cells when they are under stress.
necrosis	Death of tissue.
negative entropy	To become more ordered.
neonate	A baby who has just been born.
neovascularization	New blood vessels.
nephrotic syndrome	A kidney disease that allows albumin and protein to leak into the urine.
neurohormones	Hormones produced by nerves or regulated by nerves and include adrenal hormones, Angiotensin system and Aldactone.
NG tube	Naso-gastric tube is a tube inserted through the nose into the stomach to empty the stomach.
nitrates	A chemical compound used in explosives. Tiny amounts of this substance can cause vasodilatation and relief of angina.
nitroglycerin	A chemical compound used in explosives. Tiny amounts of this substance can cause vasodilatation and relief of angina.

Nitroprusside	A drug that causes the arterioles to dilate. Prolonged use can result in cyanide poisoning.
Non ST segment elevation infarct	A myocardial infarction with only depression, T-wave changes, or no change on the ECG. It accounts for 60% of all heart attacks. Survival is better for the first 24 hours and is worse at 6 months as compared to a ST elevation infarct.
non-ischemic cardiomyopathy	Weak, sick heart muscle due to some other cause other than coronary disease.
non-Q wave myocardial infarction	A myocardial infarction with only depression, T-wave changes, or no change on the ECG. It accounts for 60% of all heart attacks. Survival is better for the first 24 hours and is worse at 6 months as compared to a Q wave infarct. There will be no Q waves on the ECG.
Norvasc	A third generation calcium channel blocker for the treatment of hypertension and coronary spasm.
NSTEMI	See Non ST segment Elevation Infarct(myocardial infarction).
Ntg	See nitroglycerine.
Nurse's Health Study	It was established in1976 and II in 1989 to study chronic disease including vascular disease in women.
obstructive sleep apnea	Cessation of breathing while sleeping usually due to the tongue falling into the throat and obstructing the airway.
occlusions	Complete blockage with no flow passing into the distal vessel.
orphan diseases	Diseases that are so rare the treatments will never generate revenue for the drug companies and will therefore never pay for the treatment development.
oxygen saturation	The percent of hemoglobin that is carrying oxygen. 95 to 100% is a normal value.
P450 metabolism	A cytochrome enzyme in metabolism that can be over worked by some drugs.
pacemaker cells	Specialized cells that generate electricity and do so faster than other cells of the heart. They become the commander of the heart and determine the heart rate. They are influenced by adrenal hormones and vagal stimulation. The brain is the ultimate controller through the parasympathetic and sympathetic nervous system.
pacemaker defibrillator	A defibrillator that provides back up defibrillation. See defibrillator.
palpaebrae	The eyelid that is against the eye where blood vessel can be seen and pallor determined.
paresthesias	A pins and needle, ant walking on the skin feeling due to neuropathy a degeneration of sensory nerves. Diabetics have this condition and it is disagreeable and difficult to treat.

peptides	Small organic molecules proteins. They can be used to detect disease and treat disease.
perfused	Tissue that blood is able to flow through.
peripartum cardiomyopathy	An unknown disease of heart muscle that makes it weak and is associated with giving birth.
peripheral edema	Excess fluid and swelling usually in the lower extremities or dependent limbs.
pheochromocytoma	A Catecholamine producing tumor associated with flushing hypertension and is rare.
placebo	A fake pill that looks like the real thing. 25% of people will improve with a fake pill.
placebo controlled trial	A trial where a drug is paired against a sugar pill to see if the drug is better than a fake pill.
placque	A small bubble inside the artery that protrudes into the flowing blood. The shell around the bubble is a fibrous cap. See plaque
plaque	See placque - a different spelling for the same thing.
plasmapheresis	Modern day blood letting where blood is taken out of the body and the plasma is replaced in an effort to clean the blood of evil humors.
platelets	Tiny elements in the blood produced by megakaryocytes in the bone marrow. These are the first defense against bleeding. They are responsible for initiation of clotting and responsible for heart attacks.
Plavix	Clopidigril (generic name) an inhibitor of platelets by inhibiting ADP. Stents require the use of this medication and it is absolutely necessary to stay on for the prescribed time. It should not be stopped for surgery before that time. After this time Plavix has to be stopped 5 days before a planned surgery. This medicine is used to prevent strokes and heart attacks.
Plendil	A third generation calcium channel blocker for the treatment of hypertension and coronary spasm.
plieotrophic	Producing many effects.
pleuripotent	A cell that can become any cell of the body.
PMI	Point of Maximum Impulse. The point on the lateral chest wall where the apex of the heart strikes during diastole.
polygenic disorder	A disorder or disease that is a result of many genes.
polypeptide	Multiple peptides joined together to form proteins, long peptides.
post mortem	After death.
posterior leaflet	The mitral valve has an anterior and posterior leaflet. The posterior leaflet is more easily repaired.
posturing	The involuntary movement of the arms in a patient with brain injury.

PQRST	The letters of the alphabet that are used to name the various waves seen on an electrocardiogram. P-wave is Atrial contraction or depolarization, QRS-complex is ventricular contraction or depolarization, T-wave is ventricular recover or repolarization.
precordial leads	ECG leads V1 through V6 across the center and left chest.
preload	The stretch volume of the heart.
pro-arrhythmic affects	A medication that increases arrhythmias.
Proctor Harvey's stethoscope	A triple head stethoscope designed and used by the best auscultory cardiologist.
progenitor cells	Stem cells.
Proscar	A drug to shrink the prostate and grow hair.
proteomics	The study of proteins and their function.
proximal tubules	The proximal portion of the nephron involved in sodium reabsorption. This is a sub-element of the kidney.
proximal vessels	The part of the coronary artery that is closest to the blood supply the aorta.
pulmonary edema	Fluid in the interstitial air space of the lungs and is caused by heart failure. The lungs become heavy and the work of breathing increases.
pulmonary embolus	A blood clot that has gone to the lungs.
pulmonary hypertension	High blood pressure in the arteries of the lungs. The cause can be primary due to vascular disease of the lungs or secondary due to heart failure, or blood clots.
PVC	Premature Ventricular Contraction occurs when rebel cells in the ventricle take over command from the pacemaker cells of the heart. This gives a feeling of skipped beats, extra heart beats. Annoying regular irregular beats that can be felt while going to sleep and cause needless anxiety.
Q wave myocardial infarction	A heart attack that results in a new Q wave on the ECG. Anterior Myocardial infarctions will show a Q wave in the anterior leads, inferior myocardial infarction will show a Q wave in the inferior leads.
QRS	These are the letters that are assigned to waves on the ECG that represent ventricular depolarization or contraction.
QRS duration	The time of the QRS. Usually less than .12 seconds 120 milliseconds. The longer the QRS the greater the cardiac dys-synchrony.
rales and rhonchi	The noises made by the lungs when there is fluid in the interstitial space.
RALES trial	A trial of Spironolactone in heart failure class III IV that showed a great mortality benefit.
Ramipril	A blood pressure medication that is a tissue specific ACE inhibitor.
RBBB	Right Bundle Branch Block. Electrical impulses travel

from the AV node to the Right and Left Bundles to initiate contraction.

receptors Specialized protein structures that sit in the cell wall and allow chemical or biological reactions to occur when another protein or chemical structure fits into the receptor site. It is the lock that requires a key to perform a biologic function.

recalcitrant hypertension Blood pressure that remains elevated despite taking three drugs.

renal dialysis Artificial kidney to cleanse the blood of metabolic poisons.

renal failure Malfunction of the kidney so that it will not excrete metabolic poisons or remove excess fluid. A kidney dialysis machine can replace some of the function of the kidney but causes significant morbidity.

renin-angiotensin-aldosterone axis Vasoconstrictor and sodium retaining system or neuro adrenal and renal hormones.

Rennin An enzyme produced in the kidney to regulate blood pressure.

reperfusion Restoration of blood supply into tissue that previously lacked blood supply. After a blocked coronary artery is opened the heart muscle is reperfused.

reperfusion arrhythmias When blood supply is restored to myocardium that has lacked blood flow the myocardium is initially angry and can be damaged by free radicals that are generated. This results in either a slowing of the heart rate or an acceleration. At times ventricular fibrillation will ensue requiring a shock. The arrhythmias usually resolve quickly.

reserpine A old blood pressure medication that had a side effect of depression.

resident A young doctor in training who is under the supervision of senior residents, fellows, staff doctors. They have working hours that are quite long as compared to the working public.

respiratory compensation Increased respiratory rate to blow off carbon dioxide to compensate for acid production in the body.

restenosis After a balloon or stent angioplasty the vessel goes through a repair process and in 20% the vessel (coronary artery) heals in a manner that reoccludes the vessel with return of symptoms. Myocardial infarction is rare with restenosis. Medicated stents have a much lower restenosis rate and are preferred in small vessels or complicated blockage.

Reteplace A clot busting drug given in two doses and is not weight adjusted.

retrospective trials Research that looks backward in time and may have bias

revascularization	To restore blood supply by bypass or angioplasty or medications.
Reye's syndrome	A deadly illness in children preceded by a viral syndrome and is associated with the use of aspirin.
rhabdomyolysis	The breakdown of muscle fibers with muscle pain, brown urine, and kidney failure if not treated quickly with hydration.
Rimonabant	A new drug to control smoking and obesity that is still in clinical trials.
risk factor	The term coined from the Framingham study that examined various causes of heart and vascular disease. The risk factors include hypertension, smoking, cholesterol, gout, family history, and explains about 90% cardiac and vascular events.
RVU	Relative Value Unit is a work unit that determines pay for a medical service. Ultimately it is determined by the resources available and not the worth of the unit.
S3	Third heart sound which represents a distended overfilled ventricle in heart failure. In young athletic adults it can be a normal sound LUB de lub, or "Kentucky" in cadence.
saphenous vein anastomosis	A portion of the saphenous vein which runs from the foot to the thigh along the inside of the leg is surgically placed from the aorta to the coronary artery distal to a blockage. The connection between the artery and the vein is known as the anastomosis.
self titrate	A medication that slowly increases in effectiveness without changing the dose.
senescent	Old and worn out.
sinus node	The specialized pacemaker cells that command the heart and determine heart rate. It is located near the junction of the superior vena cava and the right atrium.
sodium	An alkali metal with atomic number of 11. It is highly reactive and does not exist by itself. In the body it is combined with chloride to form the ions of salt.
sodium 140 meq	The average concentration of the cation Na sodium in the body. The concentration is 140 milliequivalents per liter of fluid.
sodium and potassium channels	The regulatory proteins in cell membranes that permit sodium or potassium to enter and leave cells.
Spironolactone	A potassium sparing diuretic that decreases fibrosis and is a good medication in heart failure.
sputum	Mucous that is coughed up from the respiratory track.
ST elevation	The segment on the electrocardiogram that is between the QRS and the T wave. Elevation of this segment is seen in acute myocardial infarction. ST elevation can occur in pericarditis, early repolarization, LVH RV strain.

ST segment elevation Myocardial infarction	The segment on the electrocardiogram that is between the QRS and the T wave. Elevation of this segment is seen in acute myocardial infarction. This will result in a Q-wave myocardial infarction usually involving the right coronary artery or the left anterior descending coronary artery.
Statin	Group name for the class of drugs which inhibit HMG Co A reductase and reduce cholesterol. They have plieotrophic affects of decreasing inflammation and increasing stem cells in the circulation.
Statin myalgias	Muscle aches without muscle breakdown in 7% of patients who are on cholesterol medications the Statins.
STEMI	ST Elevation Myocardial Infarction.
stenosis	Blockage of a blood vessel or a valve. A narrowing of a structure resulting in poor function.
stent	A wire mesh that is shaped like a coronary artery. It can be smashed on a balloon and delivered to a coronary blockage by passing the balloon over a wire. When the balloon is inflated the blockage will be smashed into the wall of the artery and the lumen is maintained by the stent. The stent is a little like a tiny Chinese finger trap.
sternoclavicular	The area where the sternum and clavicle are joined.
sternocleidomastoid	A muscle that has two head and runs from the jaw and inserts into the clavicle and the sternum.
sternum	Breast bone or breast plate. It is the ridged structure that is in the center of the chest and is connected to the ribs. The sternum is sawed open to gain entry to the heart.
streptokinase	Early clot busting drug made from a by product of the bacteria streptococcus.
ST-T wave	The segment of the ECG that occurs after the QRS and includes the T wave. This segment provides information about ischemia the lack of blood to the heart.
subdural hematoma	Collection of blood between the brain and the skull.
sublingual	Under the tongue.
substernal	Under the sternum "breast bone"
supine position	Laying flat.
sympathetic nervous system	Part of the autonomic nervous system fight or flight system.
syncope	To pass out, collapse usually with some injury, simple faint
systole	Contraction phase of the heart. The time when the heart is contracting from mitral valve closure to just after aortic valve closure.
Systolic blood pressure	When blood pressure is measured it is the top or peak pressure as opposed to the trough pressure diastolic.
systolic dysfunction	Abnormally weak contraction of the heart.

systolic pulsation	A tiny tap felt over the left sternal intercostals space during myocardial contraction.
tachycardia	Fast heart rate,
telomere	An organelle of a cell that determines how many times it will divide.
Thalassemia	Inherited anemia with small cell volumes.
thallium	A nuclear isotope that can be used to show perfusion of the heart.
thrombogenic	Clot forming tendency to produce clot.
thrombolytics	Clot busters.
thrombosis	Clotted vessel.
Thromboxane A2	A substance made by platelets that causes clotting.
tissue plasminogen activator	Clot busting agent made by human endothelial cells.
titration	A gradual increase or decrease in dosage.
titration scheme	The method of increasing medication slowly so adverse reactions are avoided.
transient ischemic attack	A stroke like symptom that resolves quickly.
transplant vasculopathy	An inflammation of blood vessels that causes severe blockages and is due to a immune reaction a type of transplant rejection.
triage	To rank in severity of illness so that resources can be used efficiently for the greatest good when resources are limited.
trigeminity	Early heart beat every third beat: dah dah dah dah dah dah.
triglyceride	Fat that circulates in the blood.
Turing machine	Symbol manipulating devices that can encode secrets invented by Alan Turing.
U	The wave that follows the T wave on the electrocardiogram and may be associated with hypokalemia.
ultrasound	High frequency mechanical vibrations usually above 1 megahertz.
urethral catheter	A small tube that is inserted into the bladder to drain urine.
vacuoles	empty containers within cells that can be filled with products manufactured by the cell.
vascular repair	Repair of blood vessels.
vasculitis	Inflammation of blood vessels.
vaso vasorum	The blood supply to larger arteries that provides nutrients to the arterial wall.
vasoconstriction	Squeezing or narrowing of blood vessels usually under neurohormonal control.
vasodilatation	Dilatation or enlargement of blood vessels under neurohormonal control

vasodilator therapy	Medical treatments that cause dilation of arteries.
Vasopressin	A drug that is an antidiuretic hormone and a vascular constrictor.
vasospastic disease	Collapse of a vessel due to spasm of the muscles in the arterial wall.
venous insufficiency	Malfunction of the venous valves that result in increased swelling in the lower extremities. This can be confused as heart failure.
venous plexus	Multiple inter-connecting veins.
venous pressure	The preload of the right heart as measured by jugular vein column.
ventricle	The pumping chamber of the heart. There are two a right and left ventricle.
ventricular ectopy	Extra heart beats that originate in the ventricle.
ventricular fibrillation	Chaotic electrical activity in the ventricle with resultant death unless a shock can be delivered.
ventriculogram	X-ray moving picture of the left heart's pumping ability and will also check for mitral valve leaks.
V-HeFT I	Early vasodilator trial in heart failure.
visceral fat	The fat stored inside the abdominal cavity and is not removed by liposuction.
watermelon syncope	A Houckster term to stress the importance of diet education to prevent hyperkalemia that will result in pass out.